For Mother

Acknowledgements

The author and photographer wish to thank in particular Chris Noble, President of the Australian Hibiscus Society, and his wife Patricia Noble, of Hibiscus World, Caboolture, Queensland. They are also grateful to Rob Bayly, Joy Browne, Jack Clark, Gordon Currie, Bruce Haggo, Fane Kearney, Arno King, Jane Rumble, Dene Thomas, Bill Wilkie, Ron van Zuilen; and to all the gardeners who allowed their hibiscus to be photographed.

Page 1: The glow of red hibiscus and the suggestion of the blue sea make a stunning combination.

Page 2: Boundary fence planted with colourful mixed hybrids.

Page 3: 'Eden Rose'.

Cassell
Wellington House, 125 Strand
London WC2R 0BB
www.cassell.co.uk

Copyright © Jacqueline Walker, 1999
Copyright © David Bateman Ltd, 1999

British Library Cataloguing-in-Publication Data
A catalogue record for this book is available from the British Library

ISBN 0-304-35021-4

Cover design by Shelley Watson
Book design by Errol McLeary
Typesetting by Jazz Graphics
Printed in Hong Kong by Colorcraft Ltd

Hibiscus

Jacqueline Walker

Photographs by Gil Hanly

CASSELL

Contents

Introduction

There's nothing understated about hibiscus. Their flowers demand to be seen. Large painted petals surrounding a central staminal structure make them easily recognised. This staminal tube is typical of Malvaceae, the botanical family to which hibiscus belong, and is a dominant trait in all hibiscus – so much so that in one species, *Hibiscus schizopetalus*, the length of the elongated tube exceeds the diameter of the petals. Botanically, the structure is described as a merger of stamens and filaments fused together into a tube which arises from the base of the petals and surrounds the filiform style.

While all plant families have certain distinguishing features by which they are classified, kinship within each family is not always apparent. For example, Euphorbiaceae, a huge group that includes wafting garden perennials, tall trees and solitary desert cacti, can challenge the most astute botanist with its wide diversity. Malvaceae members are easier to identify. In addition to the central tube that is present in many, though not all, the flower formation of petals (usually five) is frequently arranged in an overlapping twist which is most conspicuous in the bud. Hibiscus flowers with this petal formation are referred to as 'windmill type'. There are other Malvaceae characteristics that the observer soon recognises, including the bark and the calyces containing the seedpods.

With more than 42 genera and more than 1000 species among its members, the Malvaceae family includes many plants of economic value, such as species that yield oil, paper and fibre, and others with medicinal properties. However, the two most cultivated are *Hibiscus*, grown today mostly ornamentally, and *Gossypium*, one of the world's most important crops, and one that has profoundly influenced political and economic history.

Gossypium produces cotton. There are several varieties, including G. *barbadense*, G. *herbaceum*, G. *tomentosum*, G. *hirsutum*, and G. *arboreum*, which is the cotton tree native to India. If you've ever wondered exactly which part of the plant yields the fluffy white stuff that's spun into thread enough to clothe nations, it's the filaments. After the flower (yellow with purple streaks and splodge at the base) withers, the leathery seedcase is left behind, and inside this, wrapped around the black seeds, is a mass of white fibres. Writing in ancient Greece in the 5th century BC, Herodotus reported that in India "certain wild trees bear wool instead of fruit that in beauty and quality exceeds that of the sheep; and the Indians make their clothing from these trees." *Gossypium* is something of a wonder plant because the seeds yield edible oil and butter substitutes, while seed residue is used for stock feed, oilcloth, putty, fertiliser, soap and nitroglycerine.

Many Malvaceae species are edible, including hibiscus. Yes, those large, often plate-sized petals are not only luscious to look at, they're also – if not exactly luscious to eat – quite edible and quite nutritious. And though they may not be epicurean in flavour, their bright colours can enliven a salad. More flavoursome and extensively grown as a food crop is *Hibiscus esculentus* (syn. *Abelmoschus esculentus*), the vegetable known as 'okra' or 'gumbo'.

Opposite: The pendulous flower of the species *Hibiscus schizopetalus* (from *schizo* meaning 'split' to describe the deeply divided petals) has an exceptionally long staminal tube in proportion to the petals.

The species *Malvaviscus arboreus*, known as 'Turk's Cap', has flowers that never fully open.

Other Malvaceae species in garden culture include *Abutilon*, the Chinese lantern or flowering maple, elegant in both leaf and flower and well adapted to mild and temperate gardens; *Alcea rosea* (syn. *Althaea rosea*), the popular hollyhock; *Malva parviflora*; *Malvastrum coromandelinum*, the false mallow; *Sida fallax*, once popularly used in Hawaiian leis; *Pavonia hastata*, the pastel flowering shrub; *Lavatera*, favourite of gardeners everywhere; and the well-known *Malvaviscus arboreus* or 'Turk's Cap'. But hibiscus are the showpiece of the Malvaceae, and the hybrids bred from hibiscus species indigenous to many countries have produced some of the most dazzling of all cultivated flowers. With more and more breeding has come more and more variety. Hybrid after hybrid has been released into the gardening marketplace so that there is now an almost limitless choice in shrub shape, height, spread, habit and foliage, as well as, of course, the main reason for growing these plants – the sumptuous blooms.

It is hybrids rather than species that are the subject of these chapters. The following advice on cultivation, pruning and maintenance, selection and landscape design applies generally to these modern hybrids. In Chapter 4, Hibiscus in Cool Climates, the cultivation of hardier species and their hybrids, together with selected close cousins of hibiscus, is considered separately.

Although hybrids hold the limelight, species have not been banished from cultivation. Many still grow and proliferate in the wild, and others continue to be grown in gardens around the world. Some important hot-climate species include: *H. brackenridgei*, a yellow flowering species endemic to Hawaii; *H. elatus*, which yields fibre from the inner bark and is still used today for wrapping cigars; *H. calyphyllus* from tropical Africa; *H. cameronii* from Madagascar, significant for its long, downward-curving staminal column and used in the breeding of many hybrids; *H. clayii*, another Hawaiian plant; *H. coccineus*, a brilliant red and large-flowered

Above: Species Hibiscus *calyphyllus*
Below left: Species Hibiscus *coccineus* a species indigenous to south-east USA.
Below right: Species Hibiscus *clayii*

grow well in the subtropics and moderately well in temperate climates, while not really flourishing in the true tropics.

In the subtropics where there's a discernible winter chill, which, though mild, is sufficient to induce a rest period followed by the spring-triggered new growth, hibiscus hybrids are at their best, for these are ideal conditions. Here, where winter moves subtly into spring without the drama of melting snow of higher latitudes, these plants endure. Their growth is slow at first, even deceptive, for the quiet greening-up of hibiscus bushes is no match for the burst of spring in cold climates. But cold climates have short seasons which finish just as hibiscus are hitting their stride. After spring, after summer, after the last cold-climate perennial has faded and shrubberies have lost their vigour, hibiscus hybrids come into their own. Calm weather with crisp nights and warm sunny days brings out their most intense colours, the best foliage, and flower quality that dazzles. Their season is long and it seems they save the best for last.

herbaceous hibiscus native to south-east USA and one of the most popularly grown species; H. youngianus, native to the marshlands of Hawaii; H. liliiflors, from the Seychelles; H. pedunclatus from Africa. H. sabdariffa, a biennial often grown as an annual for its fruits, the enlarged calyces, which make 'rosella jam', once commercially produced from crops in Queensland; H. schizopetalus, the species with the split petals and extra-long staminal tube; H. splendens, a small tree native to Australia, with exceptionally lovely flowers, widely grown and admired; and H. tiliaceus, a large tree common to all tropical and subtropical regions of the world and once the source of Pacific islands tapa cloth.

In all, the hibiscus genus comprises about 250 species – possibly more. Many are herbaceous and low ground plants, some grow into shrubs of shoulder height, and others into large trees. They are found mostly in climates that are tropical and subtropical rather than temperate. However, the hybrid progeny of these species have been bred for more cold tolerance. As a result, today's hybrids

Above left: Hibiscus sabdariffa: the calyces of this species are made into jam. Above right: Hibiscus tiliaceus, 'the cotton tree', is common to coastal regions throughout the tropical and subtropical Pacific, where it grows into a large tree. All parts of the tree have been used in traditional cultures.

Hardiness Zone Map

This map has been prepared to agree with a system of plant hardiness zones that have been accepted as an international standard and range from 1 to 12. It shows the minimum winter temperatures that can be expected on average in different regions.

In this book, where a zone number has been given at the end of a hibiscus hybrid entry, the number corresponds with a zone shown here. That number indicates the coldest areas in which the particular plant is likely to survive through an average winter. Note that these are not necessarily the areas in which it will grow best. Because the zone number refers to the minimum temperatures, a plant given

zone 7, for example, will obviously grow perfectly well in zone 8, but not in zone 6. Plants grown in a zone considerably higher than the zone with the minimum winter temperature in which they will survive might grow well, but they are likely to behave differently. Note also that some readers may find the numbers a little conservative; we felt it best to err on the side of caution.

Hibiscus grow best in hardiness zones 11 and 12, and in parts of zone 10 where there are niche microclimates. When given the special care suggested in Chapter 4 of this book, they may well thrive in the cooler parts of zone 10. In areas cooler than these zones, container culture is recommended.

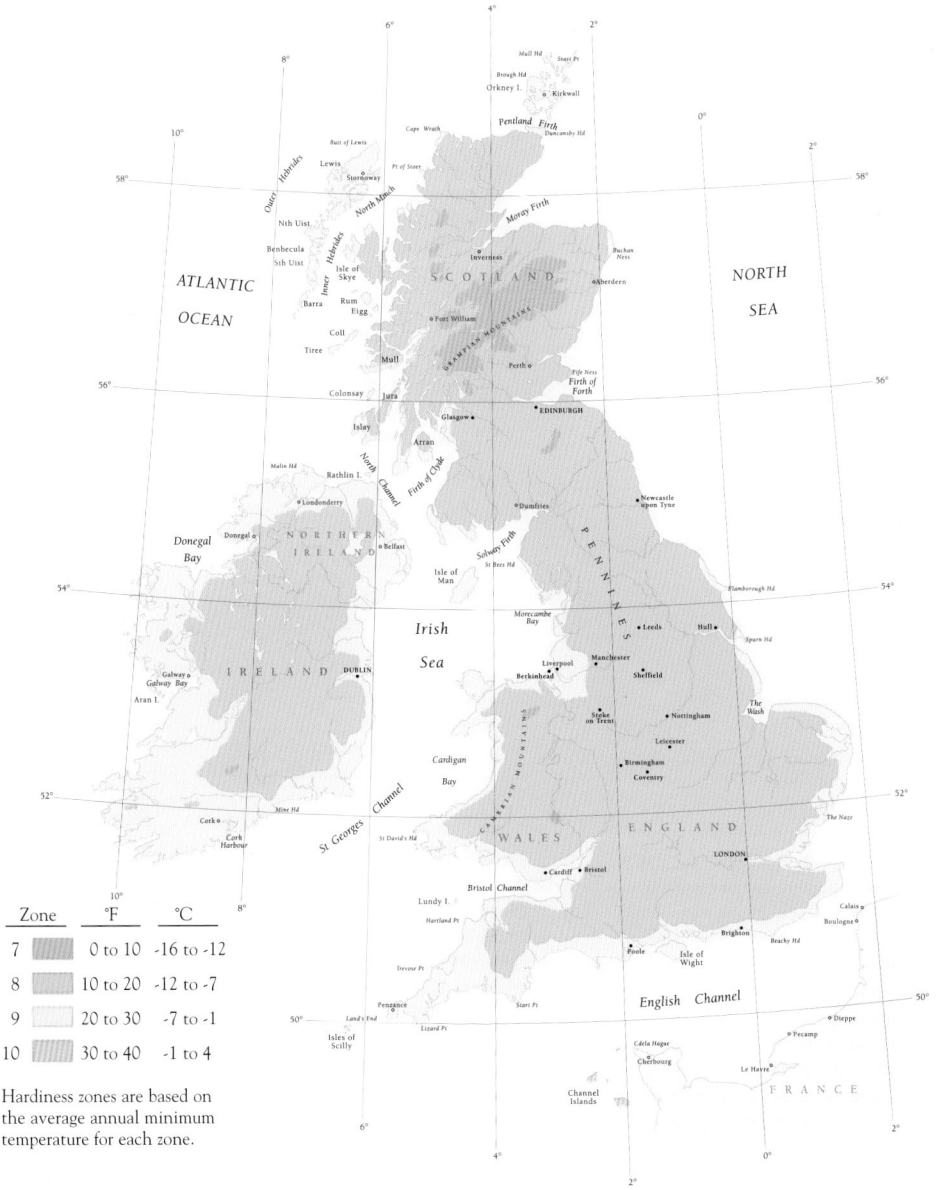

Zone	°F	°C
7	0 to 10	-16 to -12
8	10 to 20	-12 to -7
9	20 to 30	-7 to -1
10	30 to 40	-1 to 4

Hardiness zones are based on the average annual minimum temperature for each zone.

CHAPTER 1

History

If the name 'hibiscus' brings to mind palm-fringed islands, coral atolls and shaded verandas, it's not surprising, for hibiscus are coastal plants and have grown for centuries around the rim of the Indian and Pacific oceans. Colonialism in the 18th and 19th centuries helped give the flower its romantic image, an image exploited today by tourist advertising that seeks to sell holidays to oceanic destinations, to lure travellers to places of leisurely living. The name alone sells. Resorts, real-estate companies, retirement villages and restaurants use it: there's 'Hibiscus Hotel', 'Hibiscus Health Spa' … even a clothing label. (And a check on the Internet will show that the word 'hibiscus' appears in 997,940 entries.)

The name may be well known but its origins are not. Hibiscus is understood to derive from the Greek word *hibiskos*, given by the Greek physician Dioscorides in the 1st century to the marshmallow plant, a close relative of the hibiscus.

Hibiscus rosa-sinensis, an old species that was grown as an ornamental flower in China, is believed to have been cultivated there for hundreds, if not thousands, of years – so long in fact that there is no record of it being found in the wild. Although it appears to have been in cultivation throughout much of the Asian continent, early reports of *H. rosa-sinensis* flowering round temples in China imply a Chinese origin, hence the name 'sinensis'.

When this species was introduced further afield it adapted especially well in those places where other hibiscus species already grew. Everywhere the potent red-petalled flower of *H. rosa-sinensis* was seen, it

Above: *Hibiscus arnottianus* a species from Hawaii.
Opposite: *Hibiscus rosa-sinensis*, the species used extensively in hybridising.

gained a following. It was admired in the Chelsea Physic Garden in London, in Europe, India, Africa, North America … to the point that it became the celebrated flower of the tropics. The existence of other species growing in these countries enabled horticulturists to take the next step and cross *H. rosa-sinensis* with some of these. Though many of these species' flowers were less flamboyant, when genetically blended with the bold red newcomer, the results were spectacular. Thus encouraged, they tried further crossing – and behold, today's handsome hybrids were born.

Enthusiasm in Hawaii claimed a lot of attention. Using three indigenous species, including the pearl-white *H. arnottianus*, which is scented, together with

more than 33 others imported from other countries – notably the East African species *H. schizopetalus*, as well as *H. cameronii* – Hawaiian hybridists crossed and recrossed them in an extensive programme that produced a total of more than 5000 horticultural varieties. ('Horticultural' here means that the genetic parentage was recorded and the offspring identified, but not that there were 5000 varieties worthy of propagation and promotion. On the contrary: in hybridising trials only about one in every 100 or 200 will produce desirable properties of good form, foliage, flower and all-round performance.) New hybrids were created from the most successful of these – large and luscious in bloom, riveting in colour, and impressive in range. By now the hybrids had eclipsed the species, and so eagerly did Hawaiians embrace them that not only was the first hibiscus society formed there in 1911, but in 1923 a law was passed making the flower the symbol of the territory of Hawaii. Today the flower is still enshrined in law – after Hawaii joined the USA its emblem became the state flower – and it continues to delight residents as much as visitors.

Bountiful and bold, sirens of islands, hibiscus became the symbol of many Polynesian and Micronesian cultures, their flowers bedecking doorways and feasting tables as tokens of welcome, while in Indian Ocean cultures the sacred hibiscus decorated temples and religious offerings.

By the mid-20th century intensive hybridising work was being done in Florida, and the focus of attention shifted from Hawaii to south-eastern USA. Later still, Australian horticulturists began trialling new cultivars with unprecedented success. By the 1980s there were over 4000 recognised hybrids in cultivation. Today there are more than 10,000 hybrids worldwide.

Throughout all of this breeding, *H. rosa-sinensis* remained the most important genetic parent, but it was not the only species used. Confusingly, it is now difficult to establish with accuracy which hybrids are the progeny of *H. rosa-sinensis* and which are not.

Indeed, the need for clarification of the general use of the term 'rosa-sinensis hybrids' is overdue. There are Fijian hybrids and Hawaiian hybrids which are not derived from *H. rosa-sinensis* (and some which did once, but so far back in their ancestry that it's been bred out), together with many hybrids of unrecorded parentage, such as New Zealand's Jack Clark's hybrids, judiciously bred for New Zealand's temperate conditions; while the brilliant new

Australian cultivars are usually referred to as 'modern Australian hybrids'. The solution? To call them all, simply, hybrids.

The Royal Horticultural Society has formally adopted the recent proposal of the International Registrar to label them 'Hibiscus rosa-sinensis and its sexually compatible species and its cultivars'. This term covers all hybrids, together with their species parents, whether or not H. rosa-sinensis played a part

in their creation. At the same time it sets them apart from other species which have never been crossed with – i.e. have proved to be incompatible with – H. rosa-sinensis, such as the deciduous and herbaceous plants which grow in cool climates and are described in Chapter 4.

Above: Hibiscus are always at home in coastal settings.

CHAPTER 2

Cultivation

*N*ote on hardiness zones: Hibiscus grow best in the mild hardiness zones 11 and 12, and in parts of zone 10 where there are niche microclimates. When given the special care suggested in Chapter 4, they may well thrive in the cooler parts of zone 10. In areas cooler than these zones, container culture is recommended. See the map on page 11 for further information on hardiness zones.

Basic requirements for healthy growth

Hibiscus have four basic needs. These are sun and warmth, sharp drainage, ample moisture and nutrients. A further requirement, pruning, is covered in Chapter 6.

Sun and warmth

Hibiscus should be planted in the warmest part of the garden where they receive day-long sun. The cooler the climate the more important it is to give them sheltered warmth in bright sunlight. Closer to the tropics these conditions aren't as crucial, and in fact growers in these lower latitudes have found that shrubs planted in part shade, such as under the serrated umbrella of palm fronds, grow as well as or better than those in full sun.

Warmth means an absence of cold draughts, and again this is more of a challenge in cool climates where walls or fine-meshed trellises can give protection, and where raised beds with higher

Opposite: The health of these hybrids is evident in the dense, glossy foliage and the many buds ensuring flowering continuity.

edging can cosset roots, especially when mulched, while at the same time ensuring drainage. Warmth also means an absence of very low temperatures. A short cold snap in winter as low as 3°C is about the limit for these plants.

If finding a suitably sunny site is so difficult that there appears to be nowhere to plant your yearned-for hibiscus – well, go ahead and plant it in semi-shade. The foliage will be sparser and the flowers fewer, but it will certainly grow there, even if it doesn't thrive. (And don't give up. Garden conditions change. The neighbour's tall tree that overhangs your hibiscus may yet be cut back, or may die, or …?) Another solution to the lack of a sunny site is container culture, which may allow a plant to be moved around as the seasons dictate.

Drainage

Good drainage is vital. If there is the remotest chance that the roots of your hibiscus might become waterlogged at times of prolonged rain, elevate the plant. Site it on higher ground, on the top of banks and slopes, or use raised beds. Or, if this is not possible, dig out the area and build a base of rocks and gravel before backfilling. Porous soils suit hibiscus. Fast-draining volcanic soil is good; sandy loam is ideal. Coastal sites which are almost pure sand may pose other challenges, such as how to provide and retain consistent moisture and nutrients, but drainage is not one of them.

The requirement is for aeration of the roots: hibiscus need a higher than average air-filled porosity. 'Air-filled porosity' is a term used in the

17

nursery industry to indicate the percentage, by volume, of air spaces in a container of potting mix which has been watered to saturation point and then allowed to drain off. A mix containing an abundance of coarse particles will have a high air-filled porosity. As an example, roses grow best where the air-filled porosity is about 10 – 12 per cent. Epiphytic plants, which in nature grow in the forks of trees, need a growing medium of about 50 – 60 per cent. While most bedding shrubs in cultivation prefer a lower percentage, the potting mix or soil best suited to hibiscus should have an air-filled porosity of about 25 – 30 per cent.

Note that this drainage imperative applies to all hybrids and cultivated species, with the exception of *H. moscheutos* and *H. diversifolius* which are described in Chapter 4. These two species flourish in swampy soils.

Moisture

Hibiscus enjoy high humidity, but they'll grow happily in dry areas provided their roots have constant access to water, and provided their foliage is misted or regularly hosed. They need water most during the growing season. This means regular, thorough soaking. The faster the drainage, as in sandy beach gardens, the more watering will be needed. During the cold months of winter when the uptake of water is less, as it is with all garden plants, the need for watering will be less. In cooler climates with wet winters, watering will be unnecessary.

To test for moisture, examine the top few centimetres of surface soil. When it feels dry – really dry – your hibiscus needs water, and, as all experienced gardeners know, one infrequent but very thorough soaking is far better than lots of frequent but light sprinklings that may not penetrate.

Hibiscus's need for water goes hand-in-hand with the importance of mulching, which conserves moisture (see page 20), and with their requirement for regular feeding.

Nutrients

Hibiscus grown in the ground do best when planted in well-prepared sites that have friable, loamy, good-quality soil with a pH reading of about 6 – 7, and plenty of organic matter that is routinely replenished. They really flourish if, during the warm months of active growth, they're given sustained nutrition. Potassium in particular should be in their diet as it promotes the best blooms, brightest colour and most profuse flowering.

Many growers claim success with citrus fertiliser. Others testify to results with commercial rose and vegetable fertilisers, and if you have these on hand you can try them. Most nurseries use chemical nutrients in soluble form. Chemical fertilisers have their advocates, who recommend such brands as Nitraphoska, and garden-centre salespeople are always happy to advise on the use of these products. Certainly, slow-release granules are convenient for potted hibiscus, but if you really care about the environment, if you're really attuned to the ground you walk on, the air you breathe, the water you use and the water you waste, if the clear distinction between organic and inorganic is instinctive, if you're conscious of the disposal of *everything* you consume, if you eschew toxic pesticides and herbicides … then you'll bypass these packets of pricey chemicals and go organic. Even the smallest garden has room for a pair of modest compost bins, and new models have been designed for tiny, tidy backyards. And when it comes to growing edible plants in compost, the results are incontestably superior – but this is to digress: if the organic path is one you've not ventured far along, there's plenty of information at hand.

Back to hibiscus. They thrive on compost. All organic matter, including well-decomposed manure, is beneficial because it supplies nutrients to the roots, promotes a healthy balance of organisms, and adds texture to the soil. The compost you give your hibiscus should be well aged. If you're in a position to prepare your planting site several months in advance, you can use fresh manure; otherwise make sure it's well rotted. Dried manure, such as packaged sheep pellets, can be used at any time. You can also dig in decomposing leaf mould or straw, or purchased peat moss or palm peat. Mix it in thoroughly. Like all plants, hibiscus need nitrogen, calcium, phosphorus, potassium, magnesium and sulphur as well as trace elements. Good-quality soil that is regularly enriched with organic matter should contain all of these – analyses of seaweed, for example, show it to contain every nutrient known to plants, and comfrey has similar properties – although occasionally there may be a need for supplementary additives. Dolomite is useful in providing magnesium and can modify a soil that is too acidic. Borax supplies the trace element boron; potash supplies the potassium. If your soil is seriously deficient, the problem will show up not only in your hibiscus but in other plants as well, in which event it would be wise to consult a soil expert.

Maintenance feeding means lightly digging in beyond the outer limits of the shallow roots, taking care not to disturb them, as well as applying compost on the surface. Liquid feeding can also be used. How often should you feed? During the growing season hibiscus have a vigorous uptake of nutrients and can be fed as often as every five or six weeks (or four in the tropics), less frequently in the cool season. In cooler climates feeding can be lighter. Warm climates often experience forceful downpours of rain that can leach the soil of soluble nutrients, which is one reason for heavier feeding and mulching. Hand-watering, as distinct from nature's deluge, should always be done after feeding, and is essential if you use chemical fertilisers.

You can make your own liquid sprays for foliar

The flowers of this 'Southern Belle' hybrid of *H. moscheutos* are among the largest of all hibiscus blooms. As a hardy, herbaceous perennial, Southern Belle is dormant in winter.

feeding. A very dilute solution of liquid seaweed (i.e. after the seaweed itself has been removed) can be beneficial, but not every grower has access to the sea, and most home growers have to rely on their local garden retailers. There are now available a number of organic fish and marine solutions which you can buy in concentrated form and dilute for spraying. Organic gardening magazines often list similar preparations of good quality and lower prices in their classified advertisements.

Hibiscus fed little and often will be healthier than those fed heavily but infrequently.

Mulching

The value of mulching cannot be overemphasised. Mulching preserves moisture and reduces the need to water by slowing down the rate of evaporation. At the same time the loose, coarse texture of mulch aids aeration. It can also add nutrients to the soil as it breaks down, and inhibits the growth of weeds.

To some extent hibiscus are self-mulching in that they drop leaves and flowers almost year-round and these slowly decompose – another reason for never digging the soil at the base of the trunk. By all means dig a moat or shallow trench beyond the drip-line and use this both for compost feeds and for placing mulch to spread over the centre, but never dig down into the soil at the shrub's base. The area of the drip-line is the area of the roots, i.e. the root area is defined by the diameter of the outermost foliage. It is within this circle that mulch should be applied – but don't pack it right up to the stem.

So, keeping clear of the trunk's base, apply your mulch as thickly as 15 – 20 cm (6 – 8 in), although it is best to apply it more thinly, say 5 – 10 cm (2 – 4 in) and top up regularly. The weather, too, can be an indicator of how thickly and how often to mulch: in droughts, apply more.

Now to mulch materials. Use any, or a mix of, the following: pine needles, leaf mould, straw (barley and pea straw are both excellent), hay, corn husks, nut husks, chopped or shredded prunings (not infected or infested, of course), rinsed seaweed, peat products, shredded newspaper (not coloured or gloss), tree-fern fibre, coconut fibre, dried pine cones, dried bottlebrush (*Banksia*), bark (fresh pine bark should be avoided because it leaches too much resin, but small quantities can be mixed with other materials), spent mushroom compost, by-products of the sugar cane industry (as widely used in Queensland), or anything similar. Fresh lawn clippings should not be used, other than in small quantities mixed in with other materials. A very effective and stable mulch can be had from discarded doormats made of open-weave sisal/coir/jute – they're organic and make excellent aerating covers that decompose slowly.

In marginal climates hibiscus benefit from raised beds, and from the protection offered by walls and corners of buildings. (Pictured here, 'Madonna.')

Hibiscus selection, siting, planting and staking

Selection and siting

Today's hibiscus hybrids give the gardener far greater range in shape, size, habit and flower than was known a decade or so ago. Although pruning can enable you to keep a bush at whatever size you want, it makes more sense to select appropriately at the start.

Specialist growers and hibiscus society officials now recognise three sizes: tall (2 – 3 m, 6 – 10 ft), medium (1 – 2 m, 3 – 6 ft), and low (under 1 m, or 3 ft), in addition to the occasional very tall, or very low to almost prostrate. Note the use of the term 'miniature': unlike camellias, rhododendrons and other shrubs where 'miniature' describes the overall size of the bush, in hibiscus culture the description refers to flower size, not shrub size.

There is also a three-fold classification of form: open, upright or bushy. Leaves should also be considered. All hibiscus are evergreen (apart from those described in Chapter 4), and leaf variation extends from the almost circular leaf to very narrow, with many in between, while leaf margins can vary from smooth, or slightly serrated, to highly indented.

As for flower classification, this is almost a specialist subject in itself, so great is the variety that modern breeding has produced. The formal registration of hybrids and the need for uniform standards in show judging have called for some rules of conformity. Hibiscus societies now recognise four sizes of blooms: miniature (under 10 cm, or 4 in), medium (10 –15 cm, 4 – 6 in), large (15 – 20 cm, 6 – 8 in), and extra large (20 cm, or 8 in, and over). (This last category may seem very desirable to many hibiscus lovers, but because such huge blooms are man-made rather than nature-made creations, they're often top-heavy. Beginner growers especially should regard them with caution.)

Besides these four flower sizes, and besides the three categories of single, double and semi-double, there are many forms and textures. Forms include cartwheel and cartwheel overlap (twisted formation of the base of the petals – wheel-like); windmill (narrower petals separated by gaps between each); fringed petals (outer edges are split and fringed); crested (both single and double forms may have this petaloidal or small petal-like growth on the style tip); cup-and-saucer (a central tuft of petaloids arises from the centre and is distinctly separated from the outside petals); and recurved (the outer edges of the petals curve backwards). Petal textures include fluted (waves or soft ribs), ruffled, veined (contrasting colour visible as thread-like veins) and picot or frilled edges.

From the grower's point of view, the main consideration – apart from, obviously, choosing the varieties that most appeal – is the space and location for the intended plant. Consider the ultimate size of each hibiscus you buy. You are the best judge of the best plants for the best places in your garden. And when you buy, make sure you select plants that show vitality: strong stems, good healthy foliage, and signs of plentiful new growth.

Planting and transplanting

Always plant out in spring, at the start of the growing season. In mild climates it may be possible to plant from late winter through to autumn, but caution is best where weather is doubtful. The shock of a too-early planting in spring can be fatal in cool areas, as can a delayed autumn attempt when winter is fast approaching.

A young small plant can go straight into the ground untrimmed. A larger plant may need to have any long or heavy bud-bearing stems trimmed back before planting. Handle the roots gently. Invert the container or plastic bag, then cup the base of the plant in the palm of your hand to remove the container, leaving the roots and clinging soil uppermost. Observe the spread of the roots: they spread out at the sides more than down at the to the surface. Cover lightly and press gently, then water well.

If you're transplanting a long-established in-ground hibiscus to a new location, it will be necessary to prune it first by cutting back to about two-thirds, perhaps more. It should be watered very thoroughly a day or two before transplanting. Have the new site well prepared before you attempt the move. Dig around the plant's base from the outer edge of the drip-line, taking care not to cut the

centre. To accommodate this tendency, your planting hole should be wide and shallow, but well dug and prepared deeply below, and mounded gently in the middle.

Plant your hibiscus so that the join of the base of the stem is at the same level of soil as before, i.e. the hole/collar area must be clear of the surface. It's especially important that grafted plants have the graft area kept clear of soil and mulch, so don't plant deeply. Remember that the shallow roots grow close

This hybrid has the support of a decorative lattice fence.

roots; if they do extend beyond where you dig, sever them cleanly. Dig from the sides down in stages, then work your fork under the shrub and lift it carefully. A fully grown, mature hibiscus will have a correspondingly fully grown root system, so its removal and relocation will need more than one person (preferably fully grown but not necessarily mature!) to handle it. You'll need a sack or old rug under the roots for portability. Lower the shrub into its prepared new home, keeping the base of the trunk at the same level. Cover the roots, and water well.

Never feed a hibiscus immediately after planting – especially if you use chemical fertilisers – but give it a settling-in period first.

Staking

The shallow roots of hibiscus are not deeply anchored. To stake or not to stake? That is the question only the gardener can answer, because the gardener knows better than anyone else if and where the winds blow. Is the site subjected to occasional storms? Regular sea breezes? Ripping westerlies? The worst winds of all are strong cold howlers that are dry rather than moist. No hibiscus hybrid can thrive if regularly subjected to these. Certain species though, rather than hybrids, are wind tolerant (see Chapter 4).

The safest advice is to stake. And if in doubt, still stake. It's not just anchorage that's the concern. A hibiscus blown over by the wind may have its roots injured and, if damage is severe, root rot can set in. Hardwood stakes are best. Bang them into place *before planting* and use soft ties to hold the stem. Never use hard twine or wire or any material that can wound.

If you've found a spot up against a wall, or well sheltered by overhanging and surrounded by dense foliage, or in the niche of a recessed pocket, or other such special place, you may not need to stake. Or you may have trellising, or banks of boulders – it all depends on the site. Or you may be blessed with a climate that is always calm – a real rarity – so that your hibiscus need no buttressing. Obviously, because low-growing hibiscus are a lot less prone to damage, selecting plants that don't grow very tall is a wise preventive measure in windy areas.

Footnote on cultivation: These are optimum provisions, but of course it's not always possible to give each hibiscus the perfect environment. Witness those other hibiscus, those other tenacious shrubs out there in odd spots, on roadsides and cliffs, by streams and abandoned abodes, in gardens long neglected – sprawled or stunted, ragged or overgrown, they still unfurl their petals to the sun, such is their will to survive.

Raised beds ensure drainage and allow for quality controlled fill.

CHAPTER 3

Container culture

H ibiscus adapt well to growing in containers because of their shallow rooting systems. Pot plants have a place in every garden and are especially useful and versatile in confined areas. For people who move house a lot, and for apartment dwellers who garden on balconies, containers are indispensable. The portability of a containered hibiscus allows the flowering display to be seen at its best. At other times it can be positioned in a less prominent place.

Types of containers

Select containers that are as wide as they are deep, or even wider. Width is necessary because the plant's fine feeder roots spread out from the base of the stem like spokes of a wheel, almost horizontally. There are more of these roots than there are anchoring ones growing downwards. A tall narrow container is therefore unsuitable.

Plastic pots are fine, although in very warm climates they may heat up and harm the roots, and it's important to avoid one side of the pot heating up while the other shaded side stays cool. A safer arrangement in the tropics might be to put plastic pots inside outer pots. Some professional hibiscus growers recommend clay-fired pots rather than plastic because unglazed terracotta pots 'breathe'. Containers made of stone or concrete can also be used. Timber too is an excellent material for growing in a large container.

Opposite: Street landscaping — the hybrid 'Albo-Lascinatus' hybrid has been trained as a standard.

Above right: Bred for indoor culture, this 'Burnaby Series'

accommodating hibiscus if it's the moisture-durable kind, like cedar, and wooden wine barrels, if you can get them, are wonderful.

A large containered hibiscus can be heavy. If you have a yard full of potted plants and spend much of your time and back-bending strength moving them around, consider purchasing a gardener's trolley. These trolleys are made specially for carrying

25

Size

A good-sized container to start with is one with a diameter at the top of 30 – 60 cm (12 – 24 in), tapering to a little narrower at the base, with the height of the container about the same as the width. You can grow hibiscus in pots wider than higher, but not the reverse, because a tall narrow size is wasted at the bottom and too restricting at the top, so start with a 30 – 40 cm (12 – 16 in) size, and after several years repot to a larger one.

Drainage

Drainage holes should be in the outer rim of the base of the pot, rather than underneath – or both – because matted roots underneath can in time clog up drainage, and will certainly do so with pots standing in saucers – a good reason for raising pots on small blocks. Saucers have their uses in indoor culture and during times of absence when you're not there to water frequently. At such times shallow saucers can be left with water, which also helps cooling in excessively hot weather. In general though, since sharp drainage is essential, hibiscus grown in pots need sharp side drainage and regular watering.

Potting mix

Because of the need for a mix with high air-filled porosity, as explained on page 17, your potting mix should always contain some sand. The best sand to use is 'sharp' sand, which refers to the surface of the individual grains. Sand composed of sharp rather than rounded grains usually comes from granite and is better at providing air-filled porosity.

The mix must be friable. A combination of 80 per cent potting mix (usually fine pine bark) and 20 per cent sand is quite standard; or you can use a 50–50 blend of high-quality garden loam and commercial potting mix with the sand. You can add a small quantity of well-aged compost to the potting mix (making sure it has no worms in it because worms block drainage), or a small scoop of fine peat, in

containers and are sold through classified advertising in gardening magazines.

While potted hibiscus may be heavy, their growth habit is gentle. Unlike vigorously rooted plants overdue for repotting or for planting out, hibiscus roots never burst their confines. Orchids, fleshy-rooted agapanthus, shrubs, bulbous plants, and many others with expansive or exploratory roots can crack open a container, when hibiscus roots reach their limit they start to suspend growth. As they reach the pot's perimeter they turn and grow down the sides before only slowly starting to fill up the centre. You can see this for yourself if you examine one that's been potted for a few months. Shake it free of potting mix to observe the shape of the roots forming a mound over a central cavity. Only when all the outer areas of the container are occupied will the roots start to fill the middle.

In small gardens containers can be arranged according to flowering times and effective colour schemes.

addition to the sand. A soilless mix contains no trace elements, which can be supplied through slow or medium-release fertiliser. It is best if some compost is used in every mix, in small amounts only – or else applied as a surface mulch. Manure can also be used as a covering mulch.

Plant selection

Always select compact and bushy hibiscus for growing in pots, avoiding especially plants which are scraggy, or tall and leggy. Although pruning can control size, it is best to buy a smaller bush at the start, preferably one with a thick branching habit and plentiful foliage.

Planting

Fill your container about a third full and gently position the hibiscus in place, taking care not to compress or damage the roots. Don't plant too deeply. Keep the join of roots and stem close to the surface. Add the rest of the mix to fill up the pot, then lightly pat down to leave a space of about 3 – 4 cm (1 – 1½ in) at the top. Never fill right to the top. Water well. Once the plant has settled in the level may drop slightly and you can apply a thin layer of mulch. Mulching is beneficial but not essential.

Feeding and watering

If you use slow-release fertiliser in the form of coated granules or pellets, you may need to feed your hibiscus with these twice a year – once in spring and again in early autumn. In a cool climate and with small plants, once a year may be sufficient. If

'Isobel Beard', a semi-double lilac hybrid growing in a terracotta pot.

1: A newly rooted cutting, showing white, healthy roots.

2: Potted up and after several more weeks, both foliage and roots show robust growth.

3: Typical root growth. At this stage the young plant can be repotted to a larger container, although this is not yet necessary.

4: By now the roots have encased the mix in a thick mesh wrapping and are starting to fill the centre. It's time to repot.

5: Before placing the plant in the new pot, gently tease out the bottom roots to encourage them to spread.

6: When filling the new container with mix it's important to keep the stem/root join at the same depth: do not bury it deeper than the previous level. Staking is recommended.

you use a liquid foliar spray, feeding will be more often – once every two or three weeks during the growing season.

As for watering, water only when your hibiscus needs it. How do you tell? Scratch the surface to a depth of about 2 – 3 cm ($^3/_4$ – 1 in): if the potting mix is moist, don't water; if it's dry, give it a thorough hosing. Always water well after feeding.

Maintenance and pruning

How big will your potted hibiscus grow? Exactly as big as you want it to. The size of the pot controls the size of the plant. In a 30 – 60 cm (12 – 24 in) diameter (and height) container, a hibiscus will grow to a maximum of 1.5 m (5 ft) high from the base of the pot, and will be about 1 – 1.3 m (39 – 51 in) across – which is probably as large as any gardener can handle. In fact, a smaller size may be preferred.

If you do prefer a smaller bush, just prune it. Not every gardener realises that as you trim your foliage, your plant will self-prune its roots. It works like this. Suppose you reduce your shrub to about 50 cm (20 in). The reduction in foliage will trigger the retraction of the feeding roots down inside the soil. As they retract they leave tiny tunnels of air in the mix, like worm tunnels but narrower. The roots will react in this way whether it's a potted plant or a large bush growing in the ground where the spread of feeding roots extends out to the drip-line, so that the outer dimensions of the bush determines the outer limits of the roots. During this time of root reduction it's imperative that you do not feed the hibiscus. The retracting roots are not capable of taking up nutrients; they're well provided for with air from the air tunnels they leave, while the plant at this time draws upon its own store of carbohydrates. So when radical pruning results in a surge of new growth, that growth comes from the stored energy of the plant, not from nutrients in the soil.

Trimming the bush to trim the roots is a wonderfully neat system for the home gardener. Although you can repeat the procedure as a reliable way of keeping a containered hibiscus in shape for years, it will eventually need fresh medium and rejuvenation. The colour of the roots is a good indicator of general health: white roots are robust, while darker roots are often a sign of poor condition. When you repot, pot up to the next size only.

Repotting

You'll know when it's time to repot if the roots have grown through the drainage holes and the plant's appearance indicates that it has used up all the available nutrients and is ready for a fresh feed. Repot up to the next size by removing – carefully – the root-bound plant and placing it in the potting mix in the new container with the stem join at the same level. Unlike other root-bound plants that respond well to having their matted roots teased out, hibiscus tolerate only very gentle handling of their roots.

Containers in garden planning

The main advantage of container culture is of course portability, but there is also the useful option of using containers for visibility or invisibility. Garden designers suggest planting decorative or eye-catching pots with plain or unassertive foliage plants

Blue containers offer colour contrast.

that don't compete for attention, and conversely, planting plain pots with dramatic or very floriforous plants. This is a good rule to follow and it works well for hibiscus. Shrubs grown in dull pots can show off their flowers to full effect; then, in the off-season when there is only foliage and the hibiscus may look quite drab, the plain pot can be put inside a larger outer pot, one which is decorative or beautifully crafted. Interest then focuses on the container.

The strategy of hiding a plain container when the hibiscus is flowering at full throttle is just as useful. The pot can be quite invisible in the garden. Any bushy or clumping plants – annuals, perennials or shrubs, even large bulbs and grasses – can hide a pot or two. The advantages of being able to site a hibiscus in such places is the reason many gardeners grow them only in pots.

CHAPTER 4

Hibiscus in cool climates

Hybrid hibiscus

In general, hibiscus hybrids can cope with temperatures down to 3° or 4°C. In extremes, when a drop in temperature down to zero or below is short, sharp and doesn't linger, healthy hibiscus will pull through. Prolonged periods of near-zero temperatures are borderline conditions and can be fatal. But take heart – there are strategies for dealing with cold.

If you live in a climate with fiercely cold winters, the only way to keep hibiscus alive is to grow them in pots and bring them indoors until spring. Of course, the best possible set-up is a conservatory. Failing that, a light sunny area such as an enclosed porch, conservatory extension or large bay window might offer accommodation. (It's through indoor gardening that you quickly learn the advantages bare flooring has over wall-to-wall carpet. Tiled floors or rugs and mats that can be lifted are the choice of most house-plant growers.)

Where winters are a little less severe, containered plants need not be taken indoors but can survive if moved to patios adjoining the house, to covered entrance ways, garage walls, corner niches under the shelter of overhanging eaves, and other such places that provide protection from frost. This may be all that's necessary for your hibiscus to hibernate until spring; it may suffer a little from the chill (for evidence look for browning of the leaf margins, leaf drop, or, if actually frost bitten, pulpy rather than hard stems) but will still revive in the new season. And remember, young plants are the most vulnerable – the older your hibiscus, the tougher it will be.

Foliar feeding with an organic liquid made from fresh seaweed is a proven promoter of frost resistance.

Siting hybrids for winter protection

In some areas that are climatically only marginal for hibiscus, gardeners have devised ways of helping their shrubs through the winter. A New Zealand garden subjected to quite heavy frosts successfully accommodated a large hibiscus planted next to a recessed brick chimney. The fire inside the brick chimney would warm up each night (the distance between the fire grate and the hibiscus on the other side of the brick was less than a metre) while the bricks retained residual heat to release during the day. The hibiscus thrived. There's a great convenience in growing shrubs whose root systems are gentle and non-invasive. You can grow them close to drains and house foundations and need never worry.

Some growers advocate building barriers around plants growing in the ground. Barrier banks are surprisingly effective. To make them, first mix up some soil with some coarse organic material, such as shredded bark, pine needles, leaf litter or wood shavings. Then, using your hands to shape a mounded ring like a car tyre but higher and smaller, encircle your hibiscus. You can use twigs, small

Opposite: 'Southern Belle' hybrids of *Hibiscus moscheutos* will grow in climates with cold winters.

branches or stakes to strengthen the structure. The barrier can be as high as you can make it, up to 50 cm (20 in), but must be clear of the base of the trunk. This is especially important if the hibiscus is grafted, for the graft needs air circulation around it. By acting like a quilted bedcover insulating the sleeping occupant, a barrier bank can help protect the stem, the lower part of the bush, and the ground surface area from the extremes of wind and frost.

Another way of dealing with light frost is to cover your plants with hessian or sacking laid over a temporary structure. Fibrous, natural materials seem to be more effective than synthetic windbreak cloth (there's the anecdote of the sheep farmer who clothed his plants in wool clippings to successfully ride out the winter); plastic should never be used. Tall stakes can form the framework, which needs to be well clear of the foliage so that there is no contact. Of course, in areas with cold temperatures and cold wet soil, it's assumed that your hibiscus already have the advantage of raised beds. The combination of these defences – an elevated site and temporary winter cover, as well as some protective mounding round the base – can carry a healthy hibiscus through the grimmest of winters.

Even without these precautions, hibiscus will survive if only lightly touched by frost. The leaves will be burned, but in spring new growth will shoot away.

Pruning frost-damaged plants

There's a temptation to trim off frost-burned leaves immediately because they're unsightly. Don't! Never prune in winter. Never prune in cold weather. In any case, the shrub is likely to self-prune by shedding damaged leaves. Prune only when the weather has begun to warm up and there are signs of new growth. It is also advisable to feed several weeks before the pruning.

Grafted plants

Grafting takes advantage of rootstock that has proved particularly robust or disease resistant, and is used as a base for other, less vigorous or less resistant varieties. Because some rootstock has proved to be more cold tolerant than others, these have been selected for growing in cooler climates. In New Zealand, cultivars grafted onto rootstock of 'Agnes Galt' and 'Suva Queen' have been highly successful, enabling these plants to grow in areas normally considered marginal or beyond the limits of hibiscus's range. In Australia, growers have made much use of 'Albo-Lascinatus' as rootstock ('Albo-Lascinatus' is commonly but mistakenly known in Australia as 'Ruth Wilcox', an American hybrid). Gardeners in cooler climates growing hibiscus outdoors are advised to select only grafted plants. Local nurseries are the best source of advice about grafted selections.

Hardier hibiscus – other species

Some species and their hybrids are naturally cold tolerant and are ideal for areas that receive the full blast of winter, snow and all. Other species listed here are recommended, with certain restrictions, for growing in climates marginal to the subtropics, or for particular conditions, and four species (*H. trionum*, *H. esculentus*, *H. sabdariffa* and *H. manihot*) grow easily from seed as annuals and so dispense with the problem of overwintering.

Hibiscus syriacus and its hybrids

H. syriacus offers the gardener two things that *H. rosa-sinensis*-type hybrids cannot: cold tolerance and the colour blue. The flower colour of the species is closer to lilac, but among the 40 or so worthy hybrids of *H. syriacus*, from a total of several hundred, there are a few with flowers in a rich blue.

H. syriacus has long been known throughout Asia, Europe, the Middle East and the Mediterranean. There's evidence it has been cultivated since antiquity, and although research has established its origins as Asian (it's the national flower of Korea), early reports of the species growing in the Middle East resulted in the botanist Linnaeus mistakenly referring to it as 'Syrian' – hence 'syriacus'.

In cultivation it has proved to be long-lived, tough, and well adapted to dry climates and to prolonged drought. It is frequently used as a backbone shrub in flower borders, and makes an effective hedge. In form *H. syriacus* is more multiple branching, and (usually) more vase shaped than the warm-climate hybrids. The leaves are smaller and it is fully deciduous, although in the tropics it is sometimes semi-deciduous.

The flowers, sometimes likened to hollyhocks, are cup shaped, and up to 15 cm (6 in) across. Some *H. syriacus* hybrids bear their blooms like bells. There are single, semi-double and double forms, and though smaller than the giant blooms of other hibiscus hybrids, their colours more than compensate: lilac and blue, mauve and maroon, magenta, pale pink and rose-pink, violet, purple and pure white. No yellow and orange here – these are the romantic colours of the boudoir: soft, sometimes powder-puffed, and almost always with a contrasting centre of deep burgundy red.

Among recommended *H. syriacus* hybrids are: 'White Supreme' (white, semi-double), 'Ardens' (lavender-mauve, semi-double), 'Woodbridge' (wine-red, single), 'Carnation Boy' (lavender, double), 'Alba Plena' (pure white, semi-double), 'Hino Maru' (pure white, single), 'Coelestis' (violet streaked with wine, single), 'Lady Stanley' (soft pink, semi-double), 'Flying Flag' (white with red, single), 'Double Wine' (exactly that), 'Rosalinda' (purple, semi-double), 'William Smith' (pure white), 'Heidi' (violet-blue, dwarf), 'Snowdrift' (white, semi-double), 'Ruber Plenus' (maroon-red, double), 'Mimihara' (cream/crimson, semi-double), 'Diana' (pure white, single), and 'Red Heart' (white with red veins and centre, single). 'Blue Bird',

Hibiscus syriacus

developed in France with the name 'Oiseau Bleu', is the outstanding choice for a rich gentian blue with purple centre (single).

Common names for *H. syriacus* include 'Althea Rose' and 'Syrian Rose', but the most widely used is 'Rose of Sharon'. Elegant in name and in flower, its merits are many. With its preference for dry climates, it takes easily to cultivation in inland areas away from the sea (which is rare in the genus generally) while at the same time it grows well on the coast and tolerates salt air, icy, frozen winters do not bother it, and many a bush emerges each spring from a blanket of snow. It is understandably popular in North American gardens (an ancient bush growing on Nantucket Island off the north-east amputated even more. If not pruned, the shrub becomes unsightly and the flowers will be smaller. Pruning ensures health.

H. syriacus seem to grow best in the ground rather than in pots. A lack of trace elements, especially potassium and iron, has been suggested as the reason for their reluctant pot growth, in which case booster feeding of trace elements may help. With its hardiness to cold and drought, there's good reason to plant this species in the garden. It need never grow too big when controlled by pruning, and some cultivars have an ultimate height of only 1 m (39 in).

Hibiscus mutabilis

Mutable means changeable, and the Latin 'mutabilis' means undergoing change – which is exactly what the flowers of this hibiscus species do. They change colour during the course of their brief

coast of America testifies to its hardiness in conditions both snowy and saline), as it is in other continental climates which experience the extremes of heat and cold. The only climate in which it does not seem to thrive is one with year-round high humidity.

Cultivation requirements for this species are similar to those of other hibiscus. They prefer full sun, good drainage, water during the growing season, organic feeding, and their roots away from competition from other plants. They are similarly vulnerable to insect damage, particularly bud-boring beetles, but unlike hibiscus grown in mild subtropical climates where insects outlive the winter. *H. syriacus* grown in cold places outlive the insects. Like most hibiscus, they benefit from pruning, but should be pruned only in winter when bare of leaves. Cut them back resolutely by removing about half the bush; older bushes can be

Hibiscus mutabilis – The hibiscus flowers that change their colours.

life, starting off glistening white before becoming pale pink, then full rich pink, and finally, as they begin to close, rose-maroon. All these colours are present on the bush at the same time – mutant magic.

Like *H. syriacus*, *H. mutabilis* is fully hardy and deciduous, or semi-deciduous in warmer climates. The leaves are impressively large. Downy and deeply cordate, from a distance they appear heart shaped, and measure up to 14 cm (5¹/₂ in) or more in length. They're borne on spreading branches which can sometimes grow almost horizontally if given the space, although pruning can modify the spread. The common name is 'Cotton Rose' or 'Confederate Rose'. This species produces flowers in both single and double forms, of which the delightful 'Plenus', a candyfloss double, is popularly grown. Their colour change is renowned. Although the trait appears in other species and other hibiscus hybrids, nowhere is it as consistent and dramatic. Nor has this characteristic been transferred by blending with other species, since *H. mutabilis* appears to be genetically incompatible with all other hibiscus.

The novelty of the colour change is the reason for growing it. In a small garden with space for only a couple of shrubs, one 'Confederate Rose' can do the work of several. The change occurs both on and off the bush. An intriguing idea might be to invite guests to dinner and use the flowers for a table centrepiece, picking them early (the rate of change can be slowed by storing them in the fridge) and then watching the change – and the reaction of the diners – during the course of the meal.

The shrub is best grown in an open position without competition from other plants, especially without crowding from other foliage if you want the display to be seen at its best. The usual hibiscus feeding, moisture and drainage requirements apply, with feeding suspended in winter when the shrub is leafless.

Winter is the time to prune. *H. mutabilis* will grow into a small tree of 4 – 5 m (13 – 16 ft) in warm climates and if left unpruned, but a good height to maintain in a garden of modest proportions is about 2.5 – 3.5 m (8 – 11 ft). As with all hibiscus, pruning should be radical. There's a tendency for the dried fruiting capsules of *H. mutabilis* to remain after flowering and foliage have finished, so pruning by about half will rid the bush of these and ensure good growth in the following season. Hardwood cuttings taken during this winter prune can be propagated (see Chapter 8).

Hibiscus moscheutos and its hybrids

H. moscheutos is an herbaceous perennial, i.e. it dies down each winter and comes up again in the spring. Hardy to temperatures well below zero, its common name of mallow – mallow rose, swamp mallow, marsh mallow – indicates the type of conditions it favours in the wild. The name also describes the sweet confection, marshmallow. (The original ingredient that was mixed with sugar, gum arabic and egg white was the mucilaginous extract from the roots of the European marsh mallow, *Althaea officinalis*. Use of this species in medicines for coughs, burns and other ailments can be traced back to the ancient Greeks. It was this plant, not *H. moscheutos*, that the Greek physician and herbalist Dioscorides first named 'hibiskos'.)

H. moscheutos is one of several herbaceous perennial species, such as *H. militaris*, *H. coccineus* and *H. grandiflorus*, indigenous to the warm south-eastern states of North America. These have never been extensively cultivated as ornamental garden plants, but when hybridised they produced some real successes. The first hybridising took place about a century ago. It was followed in the 1950s by another programme of hybridising which built on the earlier work and culminated in the production of the brilliant series 'Southern Belle'. Today 'Southern Belle' is offered by seed merchants, along with the aptly named 'Dixie Belle', which is a dwarf form of 'Southern Belle', to gardeners around the world. Fresh seed germinates easily. It's also possible to acquire plants from cuttings if taken sufficiently early in the season to allow time for the tuberous roots to develop.

The outstanding feature of these hybrids is their flower size. On healthy, quality plants the size is staggering, for these blooms are giants, measuring as much as 30 cm (12 in) across. The colour range within the two series of hybrids includes pink, white, crimson and several reds – all with deep maroon centres. Their swamp origins give them a huge advantage for gardeners with poorly drained soil because they can tolerate wet roots in near-waterlogged conditions.

Here is an herbaceous, fully hardy, frost-tolerant perennial which will grow in situations with poor drainage, is usefully dormant in winter when you can forget all about it, shoots up in spring to grow vigorously to 1 – 2 m (3 – 6 ft), produces mammoth blooms in rich colours, and continues to do so year after year – all for the price of a packet of seeds. If these *H. moscheutos* hybrids are so marvellous, one might wonder, what's the catch? Why aren't they grown everywhere? There's no catch – only they do need a bit more care.

These hibiscus have tuberous root systems and, like all soft-stemmed perennials, their new spring shoots are tender and sappy; snails, slugs, loopers and caterpillars love them. And they must be treated like soft-stemmed perennials and protected from injury; while other hibiscus make hardwood, *H. moscheutos* are like dahlias. Their huge flowerheads make them top heavy, and they are more prone than any other hibiscus to wind damage. They *must* be staked, and they *must* be protected from strong winds. They also need more water than other hibiscus. Given their swamp origins, it's not surprising that they can take up a great deal of moisture during the growing season, but watering should be suspended in winter. Feed them over the summer, and watch for insects. Hygiene is important: any injured or damaged stems should be removed as these plants are especially vulnerable to disease. The plants should not be pruned – in late autumn, as they shed their leaves, allow them to die back at their own pace. Only when the remaining stems have wilted to a brown colour, rather than green, should you tidy up. Remaining stems can then be trimmed to a few centimetres above ground level and the area cleared of any debris which might harbour insects.

While it is true that they can keep going year after year and can be propagated quite easily from root division, some growers find that these herbaceous hybrids lose vigour after several seasons and replace them with new plants grown from fresh seed.

H. moscheutos hybrids are not for windy climates. At 2 m (6 ft), 'Southern Belle' hybrids are taller than 'Dixie Belle' (1 m, 3 ft) and have larger blooms, but both strains produce sumptuous flowers – flowers of such size that they're worth picking as a talking point. Both can be grown in pots but are much better in the ground where they can be included in mixed plantings and herbaceous borders. Massed plantings are spectacular. There is also a more recently bred strain called 'Mallow Marvels'.

Hibiscus glaber

While not fully hardy, yet more tolerant of cold than the *H. rosa-sinensis*-type hybrids, this evergreen species from Japan bears shiny, light-green leaves and single yellow flowers with dark wine centres. The flowers change to a soft satiny apricot as they age. Both flowers and foliage have appeal. The most vigorous form, *H. glaber* 'Purpurea', whose larger leaves have a dark purple sheen, is particularly handsome.

H. glaber can be grown as a small specimen tree reaching 3 – 4 m (10 – 13 ft) or more. It does not require such severe pruning as that recommended for other hibiscus, but general cultivation is otherwise the same. A useful quality is its tolerance of extreme coastal conditions. It can be grown as a hedge to buffer salt-laden winds, and is well worth growing for this purpose alone.

Hibiscus hamabo

Not to be confused with a cultivar of *H. syriacus* with the same name, this species is native to Japan and the Okinawa Islands where it still grows in the wild. It is so similar in appearance to *H. glaber* that it's been described as 'debatably a different species only distinguished by the leaf', for it shares with *H. glaber* the *H. tiliaceous*-type flower of yellow petals surrounding a purple eye and ageing to soft apricot. *H. hamabo* may be naturally semi-deciduous in the wild. In cultivation in Western gardens it has responded to cool temperatures by shedding old leaves before sprouting fresh growth in the spring. These leaves make it an eye-catching foliage plant: they're large, round and thick with white undersides, borne on a bushy shrub 1.5 – 3 m (5 – 10 ft) tall or taller. Tolerant of coastal conditions, *H. hamabo* can be pruned to a manageable size for small gardens, where it deserves to be better known and more widely grown.

Hibiscus insularis

H. insularis is another coastal species recommended for sandy sites. Its densely branching habit makes it an excellent hedge plant for seaside windbreaks. Though not for really cold climates (min. 5°C), *H. insularis* has survived surprisingly well in frosty winters of northern New Zealand, and is at least half-hardy. The species comes from a remote island off the east coast of Australia where the soil is light and sandy, the air salty, and conditions windy. It will grow in poor soil if drainage is good.

In common with many coastal plants, the leaves of this species are glossy, almost waxy. The flowers have a subtle charm unlike many of the buxom, overblown hybrids, having the 'look' of a species flower, with lemon petals tapered at the outer edges and anchored with a large dark maroon centre. They have in common with *H. glaber* the tendency to change colour (to dark gold) as they age. The shrub needs only light pruning.

Reddish leafed form of the species *H. glaber*, ('Purpurea').

Hibiscus insularis is a rare species, so rare it was once endangered, and is deserving of wider cultivation.

Four species to grow as annuals

Hibiscus trionum is an annual or biennial native to many countries. A petite, unassuming little plant, it grows a mere 50 or 60 cm (20 or 24 in), with very narrow, quite unhibiscus-like leaves and small decorative flowers. The flower's five petals, cream, yellow or sometimes white, surrounding a deep maroon centre, are borne profusely in summer. They're short-lived, opening for just the duration of the warmest part of the day and closing in mid to late afternoon.

Though inclined to be straggly, the soft-stemmed plant can make a contribution to the temperate perennial border, if mass planted. Being low growing, it needs no staking. In the sandy soils it

enjoys it may become somewhat prostrate, while in richer soils it tends to grow more upright.

In Australia, *H. trionum* is often regarded as a weed, though an indigenous one, while in New Zealand it is equally often cherished as a fast-disappearing native plant. Seed is not easily obtained, probably because the plant is so self-seeding, yet it's easy to grow and tolerant of most conditions. As an annual it lends itself to gardens with frozen winters, to rock gardens, ornamental beds, and confined spaces in urban gardens. Like the species *H. insularis*, *H. trionum* is unspectacular but underrated.

Hibiscus esculentus (syn. *Abelmoschus esculentus*) is a species grown mostly in the tropics, but will adapt to a cool climate as a summer annual. Give it the hottest spot you have (full sun), good drainage,

Hibiscus trionum, a low growing annual or biennial species.

moisture) and you'll be rewarded twofold. First, the plant (which grows to 1.8 m, 6 ft) will bear flowers of an undeniably hibiscus hue, yellow with rich dark wine centres and elongated staminal tubes; and second, if the summer and warmth last long enough, you can harvest the vegetable, okra. Okra is eaten in Asia, Africa and the USA's Deep South, where it is known as 'gumbo'. The younger the fruit is picked the better, so pick it when it's small, soft and mucilaginous, and try it in salads, or steamed and in soups. The flower buds and immature pods are also edible.

Hibiscus sabdariffa, easily identified by its bright burgundy-coloured stems, is another tropical species which can be grown as an annual to produce both flowers and fruit. Like *H. esculentus*, the flowers are yellow with a deep red basal spot, but it's a more elegant plant, with blooms that are often a light primrose-yellow or suffused with pink that contrasts colourfully with the conspicuous red stems. When flowering has finished the remaining calyces swell and turn bright red – they're edible and can be made into drinks, jams and sauces. See Appendix 3 for a jam recipe.

Hibiscus manihot (syn. *Abelmoschus manihot*), a species from China, has flowers that are very showy. Pale yellow or near-white with dark brown centres; these are borne in clusters at the ends of the branches. The very large leaves are shiny, deeply lobed and similarly handsome. The one drawback to this quite sumptuous flower – apart from prickly seed capsules – is that it shares with most other species a short lifespan of less than a day. H. *manihot* is easily grown as an annual by raising from seed sown in late winter.

Hibiscus diversifolius

This little-known and little-grown species, native to New Zealand and Pacific countries, is moderately cold tolerant. It grows to about 1.5 m (5 ft), sometimes more, and is inclined to spread. The stems are armed with small prickles and the leaves armed with hairs, while the flowers have the same pleasing colour combination of many species: pale lemon petals arranged around a dark purple-red centre. They're more bell shaped than the hybrid flowers, remaining cupped rather than opening out fully.

H. *diversifolius* is useful for certain conditions as it tolerates damp and even swampy soils.

Species *Hibiscus insularis*

Alyogyne huegelii (until recently Hibiscus huegelii but now reclassified)

A species with flower colours of exceptional beauty, this shrub is set apart from all the rest by its preference for dry conditions. Tolerant of light-to-moderate frosts, it's not for very cold climates unless the winter chill is offset by fiercely hot summers; but

Hibiscus diversifolius, a species native to New Zealand and the Pacific region, bears modest flowers that are short lived. But this species tolerates damp soils and poor drainage.

the so-called 'desert hibiscus' is included in this list because it just might fit a problem dry climate where attempts to grow other hibiscus have failed.

A. huegelii will not grow in areas of high humidity. Although its roots need available moisture when young, once mature the shrub will cope with considerable drought, and indeed will flourish in really hot dry places – which is why it grows to perfection in the Western Australian desert and in Southern California. And as for the flowers, they're delectable. The petals are arranged in an overlapped windmill format round a dark eye, and are usually lilac or purple; careful selection has produced some with deep blue flowers, hence the common name 'blue hibiscus'. In the botanical gardens south of Newport, Los Angeles, where a formerly derelict area has come to life with massed plantings of one of the bluest of these, the effect is electrifying. In Californian residential gardens too, bright *A. huegelii* flowers set against the white background of Spanish houses make a splash of blue – pure pleasure.

A. huegelii is evergreen, has lobed, toothed and somewhat hairy leaves, and benefits from rigorous pruning after flowering to keep it compact,

otherwise it tends to become straggly. It likes sandy soil, sharp drainage, full sun, and responds to light organic feeding. It is not long-lived and should be replaced every few years. Shunning humid climates, this species is the envy of many who can grow the coastal hibiscus, so if you can give it the conditions it likes, cherish it.

Hibiscus arnottianus

A half-hardy species that will grow in marginal areas that might be a touch too cold for hybrids, *H. arnottianus* comes from the mountains of the Hawaiian islands where it grows at high altitudes and hence is more cold tolerant. Its common name is 'Wilders White'. Unlike other hibiscus, this one flowers on old wood, and is not pruned to produce or enhance flowering but only to tidy and shape.

H. arnottianus also has the attribute, unique to the genus, of scent. The perfume, though slight, is discernible on the single blooms that are long-lasting, white with pale pink veins and deep pink-red staminal tubes of striking length. They're elegant flowers. However, it is not for the flowers but for the roots that horticulturists have propagated *H. arnottianus,* because while the species does have some cold tolerance, more importantly it is resistant to both root rot and borer, hence it has served as grafting understock for hybrids which lack this resistance. *H. arnottianus* has also been used extensively in hybridising programmes, with many of the marvellous cultivars of Hawaii owing their genetic heritage to this species.

This species likes high humidity, maximum sunlight, good drainage, and feeding and moisture as for other hibiscus, but not the pruning regime. If pruned, it may not flower the following season. By leaving it to grow at will, it can eventually become a small tree of 3 – 4 m (10 – 13 ft) – more in warm climates – although it is recommended that it be given a fairly severe pruning once every five years, and a light tip clipping every other year. It makes a handsome standard.

Alyogyne huegelii (formerly *Hibiscus huegelii*) has flowers in purple, mauve and occasionally quite blue, and tolerates dry climates and near-desert conditions.

Not for very cold climates, *H. arnottianus* may be worth considering in areas with borderline winters. It can be propagated by seed sown in spring or by hardwood cuttings.

Hibiscus relations for cooler climates

These are not true hibiscus but are Malvaceae members with hibiscus-like appearance.

Alcea rosea (syn. *Althaea rosea*), the enduring hollyhock, plant of the picket fence and chocolate-box paintings, is a biennial that thrives in cool climates – it's utterly, totally hardy – and is known for its spikes of cupped flowers. Easily grown in gardens with sharp drainage and full sun, and easily propagated from seed, the hollyhock is available in single and double forms in colours of the white-to-rose spectrum.

Malva moschata, the 'musk mallow', so called because the leaves emit a musky scent when crushed, is another old species in cultivation for centuries. Native to Europe, including Britain, the musk mallow grows into a metre-high branching perennial producing successive spires of mauvy white saucer flowers. 'Alba' is a white cultivar which grows to 1 m (3 ft) also. *M. moschata* is fully hardy.

Malva sylvestris is another hardy perennial, sometimes biennial. Erect to 1 m (3 ft), it bears hibiscus-like blooms in a range of colours from pure white to strong purple.

Lavatera, perhaps the most popular of all hardy Malvaceae ornamentals, includes the lovely sub-shrub *Lavatera cachemirica*, bearing silky pink flowers all summer; the perennial or annual *L. trimestris* 'Silver Cup', awash with pink flowers streaked with silver, and *L. trimestris* 'Mont Blanc', glistening with pure white pearls; but the outstanding performer is *Lavatera* 'Barnsley'. A newer, compact variety called *Lavatera* 'Lisanne' is also much admired. All are fully hardy.

Sidalcea malviflora and its cultivars are perennials with pink or rose flowers a little like the hollyhock. The clump-forming cultivar 'Jimmy Whittet' is recommended for perennial gardens. *Sidalcea* need full sun and drainage, and can tolerate the coldest of winters.

Pavonia hastata is evergreen and fully hardy. The deep maroon centres of these five-petalled flowers – in white, pink or pale mauve – make the plant very hibiscus-like. It can tolerate a little shade, likes rich, well-drained humus, and should be cut back hard in spring.

Lagunaria patersonii, the Norfolk Island hibiscus, is not fully hardy and tolerates only light frosts. In warm climates it grows into a tall pyramidal tree of 9 – 12 m (30 – 40 ft). The flowers are pale pink or rose-pink, with cream centres and five petals tapered at their outer edges. The most appealing is the rarer, deep purple variety. As a garden subject the Norfolk Island hibiscus has limitations, for the flowers are fairly small and the fast-growing tree fairly large; the main reason for growing *L. patersonii* is its tolerance of extreme coastal conditions.

Abutilon, the Chinese lantern or flowering maple, is a choice shrub. There are species and cultivars, and most are half hardy. The ornate bell flowers in white, cream, yellow, orange, apricot, pink and red hang from upright or arching stems. The leaves too are attractive. In warm climates *Abutilon* grows best in part shade, otherwise give it a sunny site well protected from wind. Staking is recommended. One species which is hardier than most and well suited to cooler climates is *Abutilon vitifolium* (syn. *Corynabutilon vitifolium*), whose luscious, pale lilac-blue flowers, large and saucer shaped, make it much admired and much sought.

Opposite: Indoor pot plants – these dwarf, purpose-bred hibiscus flower best in good light.

Indoor hybrids

Finally, if nothing else is possible, you can grow hibiscus hybrids indoors all year round. Demand for warmth-loving hybrids by people living in less hospitable climates led to some experimental tinkering, whereby hybrids were treated with hormonal growth retardant. The procedure was apparently first trialled in Canada and has since been copied in many countries, with the result that you can now buy a dwarf, fully fledged flowering hibiscus that will grow indoors. This 'new' plant resembles the outdoor shrubs in everything but an ability to mature, so it remains like a two-year-old seedling. It doesn't live long – it's not intended to, as producers can quickly sell you more. These potted specimens are released onto the market at about 30 cm (1 ft) high and wide, and grow to a maximum of about 50 – 60 cm (20 – 24 in), or to 90 cm (3 ft) if the pot is sited outside. Standardised versions are also available.

Indoors the plant will bloom year round if given maximum light. The less light it receives, the less foliage and flowers it will produce, and indoor conditions tend to produce dark leaves. Care involves the prevention of dust collecting on the leaves, for if its pores clog up the plant will sicken, so about once every two or three weeks the pot needs to be taken outside and the foliage washed down. This will also prevent infestation from pests such as mites. In general these purpose-bred plants remain disease-free. They should be fed with slow-release fertiliser and watered regularly, especially immediately after feeding, and it is recommended that they not be repotted.

Commercially referred to as the 'Burnaby' series, among other names, these indoor hibiscus flower in mostly single blooms in a variety of colours, and bring to the apartment dweller and cold-climate gardener a touch of the tropics they might otherwise be denied.

Above: 'Satu'.
Below: 'Mary Wallace'.

Above: The Chinese lantern, *Abutilon*, a hibiscus relation.
Below: 'Cinderella', also known as 'Burgundy Blush'.

Above: 'Norman Lee'.
Below: 'Jack Clark'.

Above: 'Saketani Blue', also known as 'Saketani Lavender'.
Below: 'White Fantasia', also known as 'Swan Lake'.

CHAPTER 5

Hibiscus in garden design

Bold and imaginatively planned hibiscus gardens are – regretfully – rare. Perhaps that's because there used to be a prevailing belief that hibiscus had to be planted in beds separated from the rest of the garden; beds where each bush was evenly spaced at about 1.5 m (5 ft) apart, and destined to remain the exclusive territory of hibiscus alone, forbidding entry to any other plant. Such gardens can be very boring. Blobs of uniformly spaced bushes can look a lot like civic park displays or trial beds, and not much like the kind of leafy retreat most people seek to create in their backyards.

By all means grow them this way if you wish, but do realise that it's not the only way. A fresh approach is easily gained by visiting gardens designed by professional landscapers. Professionals spend a lot of time working through the whole plan, of which plant selection is only a small part. They consider every aspect of climate and topography, of views and existing or 'borrowed' (i.e. neighbouring) landscapes, of buildings, boundaries, entrances, soil, drainage, light, shade and sun … all this in close consultation with clients. The finished design will show axis lines and focal points, include vertical and horizontal dimensions, and take into account the garden's form, depth, texture and use of colour.

All this may sound complex, but the aim is simplicity. The most successful gardens are often very simple. The secret lies in unifying materials and plants, so that instead of mixing, for example, stone, timber, paving, concrete, and brick, only one material is used throughout. It is the same with plants. Instead of planting one of this and one of that, the skilled designer selects a favourite with good form, foliage and blooms – and then plants lots of it. Repetition unifies and strengthens the structural elements and overall plan. Twenty different hibiscus in 20 different colours does not make for good design. The selection must be narrowed down. Yes, it may be difficult to choose just one hibiscus from all those delectable hybrids you've already shortlisted to a mere eight or nine (you want them all!) but there will still be a place for the others as secondary features supporting the star performer.

Planning is all-important. The smaller your garden, the more essential it is to plan on paper. Money paid for professional advice is money well spent. A tight, tiny space at the rear of an inner-city unit is exactly the sort of place that benefits most from landscaping expertise. Conversely, a large rural property may have less need of such input because it will already have areas allocated by fencing and hedges for crops, for edible gardens, pet animals, groves of natural forest, and so on. So if you garden within close confines, and especially if you're starting a new garden from scratch, consider consulting a designer. If you cannot afford one – and cannot afford to wait until you can – buy or borrow books on garden design. They can trigger ideas.

On the other hand, it should be said too that, while most gardeners delight in design – just as most people apply some sort of aesthetic rigour to choosing furnishings, or clothing, or pictures to

Opposite: Hibiscus can be used to good effect in modern landscape architecture, here used as an entrance feature. (Hybrids from rear left to front right are: sport of 'El Capitolio', 'Moonshot' and 'Golden Belle'.)

hang on walls – there will always be collectors for whom disciplined layout is irrelevant. These are the gardeners who know every species in the genus, who never turn down a chance to add and expand, who can relate anecdotes for every find, who encourage rampant self-seeding, and whose plant knowledge is always deep and often scholarly. Such gardeners and gardens have much to offer, and should be valued. But for most hibiscus growers, some guidance will be helpful.

Some landscaping suggestions

Fire and water

The most dramatic backdrop you can give hibiscus is a seascape. As coastal plants, their affinity with the sea shows to great advantage in beach gardens where they can frame a sea view. Lucky the gardener who can grow a vivid 'Big Tango', 'Red Robin', 'Mollie Cummings' or some other sublime hybrid fired from a red furnace to contrast with a sheet of blue. If you cannot command a sweep of ocean, or a chink of inlet, then even a slender gleam of silvery grey beyond the ridge can bring together these two opposing elements of energy and calm – and of course a swimming pool will have the same effect.

A solid background

A wall may be a more attainable backdrop. In cool climates the sheltering, warmth-collecting properties of a high, solid wall can make the difference between growing hibiscus and not growing them. For centuries, northern hemisphere gardeners have used south-facing walls to nurture plants that would otherwise not thrive. (Southern hemisphere walls are north facing, of course.) A wall has practical and aesthetic value. If it lacks the pleasing appearance of stone or brick – the more

Hibiscus and sea – always an appealing combination.

Above: The painted blue background, which offsets the glowing red hibiscus flowers, echoes the role of blue sea; and by limiting the hybrid choice to just one and repeating it, the result is simple but stunning.

Left: A wall is always a good backdrop. Its warmth-collecting properties benefit hibiscus while the flowers are seen to advantage.

ancient, the more the appeal – it may well be paintable. Paint can work wonders. Paint a background to display your dazzling hibiscus: perhaps olive green or dark forest green, or brown, or various lighter, neutral shades. Or you could splash out and try a rich blue. In confident hands strong purple can work magic. And in a courtyard garden, especially one with contemporary architecture, try white – in classic Grecian isle style, white never fails as a foil for potted hibiscus.

Alternatively, clothe your walls in living greenery. Self-clinging vines can cover a fence or wall so densely that the underlying structure is invisible. Ivy is an obvious example, although it's

As a shrub grown almost exclusively for its flowers, it is unsurprising that its use as a foliage bush has been ignored. Yet within the huge range of hybrids that offer every kind of leaf size, shape and texture, as well as form and habit of branching, there are some with such handsome foliage that they can be grown for this alone. Regular clipping enhances and encourages the foliage quality. The cultivar 'Andersonii' is one of the best for this purpose. It's a tough, tolerant grower with small scarlet blooms that in a sprawling, unpruned bush tend to be pendulous, but on a clipped hedge are borne close to the foliage, which has the additional attraction of producing fresh new growth with a bronze tinge.

So consider growing a hibiscus hedge. It will make a bold framework, as well as providing a perfectly matched backdrop for other hibiscus. However, discipline will be needed. A short, half-hearted hedge is not worth bothering about. Make it long and strong or not at all. It will also take discipline to shear it regularly of all those attractive

not to everyone's liking and can be too tenacious. *Ficus pumila* (zone 10) is as dense as ivy and a better alternative. The rich flowers of *Campsis radicans* (zone 8) match the fiery colour range of hibiscus so the self-clinging vine can be both backdrop and colour coordinator. Climbers that sprawl and send out long arms and tentacles are not recommended, but there are many others that are less obstructive. *Senecio macroglossus*, the wax vine (zone 10), needs some control but is not rampant; *Hoya* species are gentle climbers and trailers, although they prefer shade; *Manettia luteo rubra* (zone 10) from South America is very moderate in habit, *Rhodochiton atrosanguineus* (zone 10) even more so; *Parthenocissus quinquefolia* (zone 8) is the self-clinging virginia creeper famed for its deciduous foliage; *Schizophragma* (zone 8) is also self-clinging; and the hardy silverlace vine *Polygonum aubertii* (zone 8) is of easy culture. Other choices might include the bluebell creeper, *Sollya heterophylla* (zone 9), and the charming black-eyed susan, *Thunbergia alata* (zone 10).

Worries about the roots of climbers robbing nutrients from hibiscus territory can be solved by planting the vine on the other side of the wall or fence and encouraged to grow to the top and spread over the sunny side. In any case, most vines send down deeper roots than the shallow-feeding ones of hibiscus. Except in cold climates, upwardly mobile plants do best with their base in the shade and their upper levels in sun, so by positioning them on the cool side of the wall both climber and hibiscus benefit.

Hedges

Hibiscus hedges are undervalued as structural components in garden design.

Opposite top: These extra tall container-grown standards have been grafted twice: first, onto other rootstock, then the 'head' onto the stem. But this treatment is not needed to grow a successful standard.
Bottom: A newly clipped hedge of 'Andersonii'.
Right: Albo-Lascinatus grown over an archway.

flowers (though you can collect them up and fill a bowl) but your efforts will be well rewarded.

To plant a low hedge the spacing needs to be about 70 or 80 cm (27 or 31 in) apart. A high hedge, up to 2.5 m (8 ft) high, will need a spacing of about 1 – 1.3 m (3 – 4 ft), and a higher hedge even more. Ensure that the individual plants are of the same size and in good health before you plant; rather than dig and prepare separate holes for each plant, it is better to dig a trench for extra drainage and for filling with extra compost.

Pleached hibiscus hedges

A variation on hedge-growing is pleaching, which means intertwining by encouraging branches of closely spaced trees or shrubs to grow together. The word comes from the Old French *plessier*, meaning to twist or to plait. When combined with trimming, the method produces a hedge with an intricate branch structure on a free-standing row of clear trunks. Horticulturally, it has a long history dating from Tudor England when pleached avenues were a status symbol indicating the number of gardeners a landowner could afford; artistically, it is a classic design form widely admired.

To pleach hibiscus a strong and well-constructed framework is needed until the row is fully established. The main staking posts for each trunk must be driven into the ground before planting, and they must be as high as the intended finished hedge. Spacing between the posts will be about 1.7 – 2 m (5 – 6 ft). Horizontal wires, tightly stretched, form the framework to which the lateral branches are tied. Laterals are encouraged to grow in a flat plane, while any wayward shoots appearing on the trunk or elsewhere are pinched out. Once a densely woven screen has been established it can be clipped and maintained like an ordinary hedge.

Certain hybrids make better pleached hedges than others. Those which make good standards are best. The cultivar popularly called 'Ruth Wilcox', correctly named 'Albo-Lascinatus', grows well when pleached, or when grown as an overhead arch or an arbour. To a lesser extent the species *H. schizopetalus* can be used this way. Both *H. schizopetalus* and 'Albo-Lascinatus' are less hardy than the reliable 'Andersonii' mentioned above.

Other foliage uses

For a foliage-only shrub that is visually arresting, try growing the cultivar with variegation, 'Snow Queen'. At a distance it shimmers. Close up, the ice-white leaves show much more white than green, making a splendid foil for the scarlet blooms (although these flowers are not the reason for growing this variety). Give it a position of prominence, such as a stand-alone bush – kept well shaped by pruning – maybe at the far end of a path or walkway. If possible, site it in part shade, for filtered sun heightens the effect.

'Cooperii' is an earlier hybrid with variegated foliage, but its performance is surpassed by 'Snow Queen'. Another, with tri-coloured foliage in green, white and pink, is 'Rose Flake'. For a purple-foliaged effect, try growing the species *H. acetosella*, known as the 'false rosella plant', a species with remarkably plum-coloured leaves and plum-coloured flowers to match. *H. glaber* 'Purpurea' has similarly tinted foliage, while the hybrid 'Andersonii' bears leaves with a bronze-red sheen.

Standards

Unless you're an experienced grower, it is probably more convenient, and certainly faster, to buy a ready-made standardised hibiscus than to create your own. But standards come at a price. The high cost reflects the time that went into the process of training a straight stem upwards to be topped off with a crown of foliage, as well as the expertise that went into the grafting of that stem, although not all standards are necessarily grafted. Hybrids that make good standards on their own roots include 'Pink Psyche', 'Albo-Lascinatus', 'Sprinkle Rain', 'Sylvia Goodman', and the white form of 'Fantasia'.

To grow your own, you'll need a clump of young

Left: 'Snow Queen' – a hybrid with the unusual combination of bright red blooms borne on white variegated foliage.

plants packed closely together in enough shade to make them head up in search of light. This takes time. As they grow, monitor them and remove side growth as soon as it appears, and keep them staked. Select the most vigorous of them as your specimen standard, but keep them together to prevent branching. Aim for a final height of 1.5 m (5 ft), possibly taller, and then encourage branching and leaf growth by giving them space and sunlight. A good shape is obtained by very frequent light clipping, which is kept up until the 'head' has reached the shape and size you want. Some designers claim that the ideal proportion to aim for is a height and width measurement that is half the height of the stem.

If you're not concerned with cost, or with the challenge of growing your own, you can ask a

Above: A bright red hibiscus adds a burst of colour to this lush subtropical garden.

Espaliered hibiscus

Generally, hibiscus do not lend themselves to being espaliered in the classic manner of those ancient and gnarled pear trees growing against equally ancient walls in Northern Europe. Pear and apple branches frequently grow at right angles and, compared with hibiscus, will 'set' themselves this way for ever. In contrast, hibiscus stems remain pliable, so espaliering them may be only partially successful. You'll need a fairly straight main-stemmed variety with strong side shoots, and you'll need wires and tying twine, and plenty of patience. The pliable stems allow them to be bent without snapping so that in time they should stay where you want them to. It's not for beginners.

Hibiscus in garden beds

The advice given in the past to plant evenly spaced hibiscus bushes in their own beds undoubtedly benefited the hibiscus, but may not have benefited the home owner. Not only is such expansive sunny space rare in the contemporary urban garden that often serves to provide privacy and seating/ entertaining areas in a narrow plot at the rear of the house, but it's hardly an inspired design. The aesthetics of design need to be worked through in conjunction with the garden's functions.

If open space is limited to a central area or courtyard assigned for seating, the surrounding area will be most efficiently and effectively utilised if it is raised. Terracing and step-down levels can be ideal for hibiscus, for several reasons. A series of raised levels that at the highest might be 1 to 1.5 m (3 – 5 ft) above the original ground level will, firstly, allow the plants growing there to receive more sun; and secondly, guarantee good drainage. It will also enable you to get the best soil mix at the start and be more accessible for maintenance. By planting at different levels, closer spacing is possible, and visually, the vertical view, with flower and texture interest at varying levels (that means the least attractive part of hibiscus bushes, the base, will be

specialist grower to custom-make you a standard of your choice, for there are hundreds of hybrids that can be made into respectable standards. The selection at retail garden centres will be more limited.

Once you've got your standard, take care with its placement in the garden: make it a focal point. Be wary of windy sites because the stake to which the standard is tied may be thinner than normal, to keep it from being conspicuous. If you plan to grow your standard in-ground, you'll need to have prepared the site in advance. If you keep it in its container (and often such a quality plant is packaged in a quality container), take your time experimenting with the site so that it will give maximum impact. Visual impact is, after all, the reason for sculpturing plants. Feed and water it as you would any hibiscus, but give closer attention to keeping it in its intended shape.

The lower branches of this mature hibiscus have been removed to keep it from cluttering the picket fence.

make successful companion plants, as well as having

Epiphytes hold little threat for hibiscus and can 'Purple Heart.'

garlic *Tulbaghia violacea*, and *Tradescantia pallida* 'Purple Heart.'

Ophiopogon, especially the all-tolerant *O. japonicus*, and *O. planiscapus* 'Nigrescens', also the flowering garlic *Tulbaghia violacea*, and *Tradescantia pallida* include the Australian native *Viola hederacea*, *Ajuga*, attractive than bare soil or mulch. Suitable plants water. And of course, ground covers are a lot more wilt in time to alert the gardener of the need to indicators of moisture levels because they'll start to During times of drought they're also useful covers help maintain a consistent acidic level. soil and as a temperature buffer for hibiscus, ground weed-free. As well as an aid in stabilising the surface living mulch, helping to keep the ground moist and

Ground-covering plants can fulfil the role of a or mounds.

plant your hibiscus on minimally raised edging shrubs can always be grown close by hibiscus if you tidy root balls, are ideal, and small, non-invading are obviously excluded. Palms, though, with their suggested below, or that go much deeper. Most trees than the surface mulch, such as the ground covers plants whose roots either don't penetrate deeper ornamentals with hibiscus, the trick is to select

If you garden on one level and want to mix other

Companion plants

hibiscus need only be 38 – 40 cm (15 in) diameter. raised beds. Remember that the planting area for fixtures, so that they themselves become miniature containers and partially bury them as permanent level. Better still, remove the bottoms of the in) or so by placing large containers on the higher modest scale, height can be gained at only 50 cm (20 natural large rocks all make good structures. On a them), timber edging, logs, and of course stone and Blocks, bricks, railway sleepers (if you can get elevated planting screens neighbours, too.

and enhancement to the whole garden. Such concealed by surrounding shrubbery), gives structure

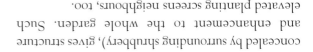

Above: Colour and texture – while the hibiscus colours blend beautifully with the paving below and the brick wall behind, there is striking textural contrast with the succulents in the foreground.

Below: An unusual combination of vertical conifer and hibiscus is unified by the white background wall.

large nor very fine.

choosing mid-textured foliage that is neither very that doesn't detract should be used. This means blended colours may be more successful – and foliage be chosen to blend or contrast with hibiscus – then discarded at the end of the season. Colours can or put in as seedlings can be fed and cared for – and for the long term. Massed plantings grown from seed with anything temporary, and annuals are not there are annuals. Hibiscus grow happily in close company

Perhaps the most reliable of companion plants receive filtered sunlight.

pockets of small logs and stumps positioned to tolerance. Orchids, too, might find a home in hibiscus may need to be selected for sun rainforest appeal. Bromeliads planted around

Colour

The hibiscus colour spectrum widens with every new hybrid released into cultivation, but it's strongest in the yellow-orange-red-pink portions, thinning out at the purple-lilac edge, with two marked exceptions: the species *Alyogyne huegelii*, formerly *Hibiscus huegelii*, of which there are forms in deep blue; and *H. syriacus* and its hybrids, which are almost exclusively in the purple-blue range. Of course, there are plenty of white and pale 'pastel' hybrids.

Strong colours are best grouped together. Clear yellow and yellow-orange hibiscus go well with marigolds. (Don't dismiss familiar faces for their commonness; undemanding culture frees up your time to spend on hibiscus and other plants.)

Hibiscus and palms bring to the entrance of this cottage-style house and trellis arch an unexpectedly tropical appearance. The soft lemon hibiscus on the right is especially pleasing when surrounded with foliage plants.

Sunflowers, *Zinnia*, *Tagetes*, *Calendula*, and *Rudbeckia* come in these shades, as does the gleaming *Eschscholzia* (Californian poppy). Red, apricot and pink hibiscus can be similarly matched. Nasturtiums are good for attracting aphids away from hibiscus, as well as for their hibiscus colours. Perennials include *Geum*, *Gerbera*, *Anigozanthos*, *Kniphofia*, and the outstanding *Hemerocallis* (daylily) whose galaxy of new cultivars allows you to colour-match your hibiscus with precision.

When thinking about colour, don't confine yourself to flowers but give as much, or even more, attention to foliage. Ivy-leaf geranium, for example, can be a better way of tying colours together in a hibiscus garden, especially in wine-red shades. And look also beyond plants. Look to the colours already existing in the garden by assessing (the merits or otherwise of) the colours of containers, statues and ornaments, seats, tables, fences, walls, tool sheds and utility areas, gates ... all of these should be considered as part of your canvas.

It is regrettably true that fashion frequently dictates not only to the retail market but to gardeners' plant-buying decisions – however firmly resolved the will to resist – and flower colours are a good example. When Vita Sackville-West created her white and silver garden at Sissinghurst it was a stroke of genius, but all those copies that sprouted in the 1970s eventually became tiresome, which is why strong colours took hold in the 1980s and 1990s. Perhaps now is the time to revisit white. Perhaps a fresh version, a white and cream garden with a

The choice of colour of both painted background and of hibiscus are clearly intentional – and very successful.

...spotlighted hibiscus. Gardens with drier, succulent landscapes need careful planning if the dry plants are to coexist with moisture-loving hibiscus, but it can be done. Hibiscus in containers, or embedded semi-containers kept separate from the succulents, can be treated separately, and moisture retention can be aided by thick mulching.

Hibiscus flowers are not to be underrated. A healthy, vigorous hibiscus — which is what they all should be — proclaims its presence boldly. So however you use them, give them the space and the light they need and they, in their turn, will illuminate your garden.

Themes

Whatever the theme of the garden, the extent of its success is the extent to which it is carried through: that means resisting the urge to go off in another direction and plant other things that are not part of the plan. Hibiscus are versatile and can adapt to all sorts of treatment. The smaller-flowered, denser-foliaged ones lend themselves more to shaping and to the standards associated with formal gardens, while the big blowsy blooms of some of the double multi-hued hibiscus belong to a less structured, highly floriferous theme. Rainforest themes are achieved with an emphasis on foliage, where waterfalls of green provide the setting for a few, tropical theme centred around hibiscus, could lead the way to new ideas. A few luminous blooms of 'The Bride', 'Great White', or 'White Picardy' might be surrounded with mass plantings of that white-leafed hibiscus 'Snow Queen', grown strictly for the foliage, and contrasted with large-leafed, strongly formed tropical foliage plants, together with white, night-scented blooms of the rainforest. Perhaps the red flowers of 'Snow Queen' could lead to a red and white theme. And another idea: fiery orange-red on a green background could be achieved by planting masses and masses of foliage plants with large leaves of high lustre and setting against this the ember glow of a hybrid such as 'Fire Engine' or 'Big Tango'. The temptation to include other species or even similar hibiscus varieties would have to be resisted, but in compensation would be the freedom to splash out on a host of foliage plants. There's plenty of scope for experiment.

Opposite: Hibiscus in small places: a narrow paved pathway, lushly planted on both sides, seems more like a rainforest than a confined inner-city garden.

Right: A number of elements are used in this contemporary design: curves contrasted with the perpendicular palm, rounded foliage, dark greens against pale walls, and the pleasing effects of a pastel palette.

CHAPTER 6

Pruning and year-round maintenance

Pruning

Left unpruned, all hibiscus soon become shapeless and messy, and their flowers fewer, smaller and duller. Most hibiscus are evergreen. Apart from a cluster of species classified as herbaceous, which includes smaller annual or biennial sub-shrubs found growing in the wild, the only two fully deciduous species in cultivation are *H. syriacus* and *H. mutabilis*. These two hardy shrubs require heavy pruning to perform well, for like their *H. rosa-sinensis* relations, they flower on new wood. Prune these two species in winter only, when the branches are bare. Other hibiscus should never be pruned in winter.

Early spring – at the start of the growing season – is the time to prune your hybrid hibiscus. Occasionally, there may also be a need for a mid-season trim when the plant is in active growth, especially in warm climates. Most experienced growers prune just as cooler weather gives way to warm and when observation reveals the swelling of new growth buds. Don't wait for these buds to open out fully into leaves, but strike just before. Such timing will give you control over your plant by directing it where to put its energy. If you wait until after the new foliage has shot away, you've already lost some control, for the trimmings you discard are the wasteful product of the plant's energy reserves which might better have gone into leaves and blooms where you want them.

Opposite: A well shaped and maintained shrub of 'Samoan Orange' in a street setting.

It's likely that along with the new foliage growth, you'll also be sacrificing flower buds. Don't think of this as a loss. Remember that you'll get far more flowers from pruning, because from each pruned branch or stem three new stems will grow in its place.

The purpose of pruning is to improve the shape of your hibiscus and to improve the vigour of its main branches (upright or spreading, depending on the hybrid) while also opening up the centre to air circulation and sunlight, thereby encouraging new blooms. Pruning is more than just reducing the length of the branch and the number of its side shoots, for it also involves cleaning up any dead wood, removing weak or bug-eaten branches – and any diseased ones, of course – and cutting away any misshapen, spindly, odd-angled or aesthetically displeasing branches. Pruning means attaining desired shapes, and desired shapes can include espaliered and standard forms (see Chapter 5). If you prune wisely and your hibiscus are well nourished, they will repay you with abundant and bountiful blooms.

By cutting back previous years' hardwood to reduce size, and by removing unwanted stems to induce vigour and shape, the result is likely to be a bush about two-thirds the pre-trimmed size. A healthy plant will tolerate even harder pruning to half size, or you may wish to take away only an eighth, or a quarter if the plant is young. With young shrubs, and sometimes even with larger ones, some experienced growers prefer to pinch out the first new

This narrow bed of hybrids is kept low with pruning.

growth rather than prune a longer branch, claiming that this method results in a thick-textured, better shape; they save the radical one-third-off prune for mature specimens.

Now for the actual cutting part. Make sure your secateurs and lopping shears are sharp and scrupulously clean, so that your cuts are likewise sharp and clean. Cut at a slight slant as with all hardwood, about a 45° angle, with the highest part of the slant just above a bud or 'eye'. This will allow water to run off rather than collect on the wound. A jagged cut will invite bacterial infection.

To remove unwanted stems growing at angles from the framework, cut at a fraction beyond the join. This applies also to those very low branches that sweep the ground near the base of the trunk. These branches are often never allowed to grow. While some gardeners like the appearance of bushiness right to the ground, others argue for their removal for reasons of hygiene, and say that space at ground level makes mulching and maintenance easier.

Opinion is also divided about the use of pruning paste. Sealing over the cut immediately following surgery is believed to prevent infection and promote health, but many arborists now eschew this practice with forest trees in favour of allowing the tree to self-seal its wound. Self-sealing occurs, they say, when the cut is made further out on the limb from where it joins the trunk, so that when this amputated stump naturally rots away, the wound left on the trunk is smooth. Hibiscus growers will have to make up their own minds, but with ever-present threats of garden pests and diseases, they may feel some security in keeping at hand a cautionary tube of paste.

Neglected shrubs that have matured into quite hefty trees need a bold approach. Firstly, feed and water them well over the growing season to promote health, then, after a winter rest period, arm yourself with a strong, well-sharpened pruning saw (and arm yourself with strong resolve, too), and lop down the entire tree to about table height, retaining half a dozen, at most, strong branches which open outwards, vase shaped, while removing inner-angled branches which clutter the centre. Clean up completely the ground below. Now, with more than half your bush or tree gone, together with by far the greater portion of the foliage, resist the urge to feed it. There is insufficient foliage left to take up and release moisture, and adding nutrients, especially in concentrated form, can harm the bush, so leave it to recover. Recovery may take another whole season, but in time it will flourish fully. The foliage and flowering will both be invigorated.

At the other extreme, in the very small garden that is perhaps a mere extension of an apartment, pruning may need to be light but more frequent during the growing season to keep a small bush compact. The purpose-bred, tub-sized hybrids that are now available make it possible to fill a miniature garden with a number of continuously flowering hibiscus of modest dimensions, and they can be kept that size by pruning.

Pruning goes hand in hand with transplanting, planting out and potting up, feeding, watering and mulching, so this is the time to consider a maintenance checklist for year-round hibiscus culture.

Maintenance in the hibiscus garden

At the start of spring
Prune. In cool climates, potted hibiscus that have overwintered under shelter can be brought into the garden. All plants, whether in the ground or in pots, should be fed at this time.

As the season advances
Apply a dressing of compost and, as temperatures start to rise, mulch. Mid-spring is a good time to relocate any plants and to plant out new ones. If pruning was earlier overlooked, it can be done now.

At the start of summer
Regular watering may be necessary, depending on rainfall. There will be plenty of fresh foliage and the first full flush of flowers appearing (later in cooler climates). Weed. Watch for insects: whenever possible pick them off, squash them, or hose them away. Serious infestation may need more serious treatment.

Mid-summer
Increased heat will bring increased pests, and

spraying may be necessary. Mulch can be topped up. A light but frequently applied layer is better than a heavy infrequent one. Feed and water. This is likely to be the time of greatest demand for moisture. Occasional mid-summer pruning can be done, or may be necessary, on vigorous shrubs. This is the time for taking cuttings for propagating.

Late summer to early autumn
Keep watering and topping up mulch, tapering off as autumn draws to a close. Check for general health, and watch in particular for insect infestation. Flowering should be at its best, and blooms longer-lasting. In cooler climates this will be the time of the last feed of the season.

In late autumn
Gather your last hibiscus blooms (along with your last tomatoes) and decorate your dining table. Only in warm areas should you continue to feed the plant. In cool climates prepare protective measures for ground-grown hibiscus, and move potted ones to shelter for the winter.

CHAPTER 7

Pests, diseases, and solutions to problems

Everyone knows the cliché, 'the answer lies in the soil', and the claim that strong, healthy plants are disease resistant, and the familiar refrain about 'prevention being better than cure' … they are all true. All are applicable to hibiscus culture. So look to the soil, the siting, the moisture, the nutrients and the care, and ensure they're of top quality. Problems should then be minimal.

The organic philosophy

The aim of all organic practices – permaculture, biodynamics, natural farming – is not just to reduce and ideally to eliminate the use of toxic chemicals that are harmful to humans, animals, plants and the planet, but to understand the processes of nature in order to participate, rather than to interfere.

This means a change in thinking. Instead of thinking monoculture on a massive scale, think biodiversity over a smaller area. Instead of thinking of insects as pests (remembering that pesticide manufacturers are powerful and influential), think of them as part of the natural order, with the plant-damaging insects kept in check by predators. Instead of thinking of perfection in ornamental and edible plants as blemish-free (sales conditioning), know that nature is not driven by glossy advertising. Think always of the way plants live without interference from humans.

Opposite: Plant health is evident in this residential garden where foliage and flowers are vigorous and buds numerous. ('Isobel Beard' is in the foreground.)

Food for resistance

If you feed your hibiscus with compost – good-quality, balanced compost – you can never give them too much. Conversely, if you feed any plant with chemical fertilisers, you can easily give them too much. Too much can kill off all microlife. With edible plants, too much can contribute to cancer-promoting nitrites, and the imbalance that can result from overfeeding with artificial fertilisers can create pest and disease problems. Nitrogenous fertilisers in particular can be harmful because they promote rapid soft growth that attracts sucking insects.

Just as mammalian predators pick off the runt of the herd, so insects will overcome the weakest of the crop. If you doubt that compost is resistance food, try an experiment using both organic and inorganic fertilisers by growing carrots or cabbage in two lots. Give one lot the best compost you can and the other lot a nitrogenous fertiliser, such as sulphate of ammonia. The latter will quickly be eaten by pests.

Another analogy is stress: stress in humans results in vulnerability to colds, flu and infectious diseases. Stress in plants makes them similarly susceptible.

A soil rich in organic matter not only provides nutrients but inhibits the growth of fungal and bacterial diseases, such as *phytophthora*, by encouraging the antibiotic activities of mycorrhizal fungi. And of course, the texture of such a medium provides hibiscus roots with that high air-filled porosity they need. When they are healthy, well-fed hibiscus shrug off or outgrow pest damage.

Stability in the garden

Aim to provide all plants with the conditions they need – if you can't, don't grow them. Aim for both lateral and vertical diversity, by matching plants with the conditions that suit them at every level, from ground-covering creepers through mid-high shrubs to the top of the canopy.

A stable garden does not contain impossibly difficult plants from quite different climates and conditions. Nor does it grow a huge number of similar plants in isolation from all others. And apart from the procedure of planting and transplanting, a stable garden is not routinely dug or hoed. Rather, layers are allowed to build up in the way a rainforest floor accumulates. Mulching and layering duplicate the processes of nature.

A stable garden in which natural pest control is practised will always have a constant, but manageable, supply of pests providing food for a constant number of predators. Both are accepted as part of the ecosystem. Not all insects are bad – some are predators, some never cause plant damage, and others never build up their numbers.

Companion planting for insect control

Companion planting is a component of the stable garden, but it is not a cure-all. On its own it is ineffectual: it doesn't rid the garden of pests, but it does aid diversity and stability. Through releasing different scents, through utilising different levels of soil, and by occupying different garden spaces in certain combinations, companion planting can attract, or repel, specific insects. For example, plants which attract aphids and are good companions for hibiscus include the annuals feverfew, coriander and nasturtiums, and the perennial cedronella.

Hibiscus pests

Some hibiscus pests are worse than others. The larval grubs of garden butterflies can chomp their way through many a hibiscus leaf, but their harm is mostly visual. There are grubs which live off the foliage of shrubs and soft-wooded trees, such as the green looper, so called because it moves by looping up its rear to its front, and other grubs which usually hatch out simultaneously. This means that if you find one, you'll find more. Watch closely for them, especially in the early morning, and if you discover them in the early stage, you may be able to pick off and destroy the whole batch.

Chewing insects

Beetles too can be picked off. Apart from the flea-like bronze beetle, which leaps too quickly to be caught, most beetles can be hand removed if the bush is not first disturbed.

Snails and slugs, though voracious chewers, are a threat only to fresh new leaves, mainly on young plants. The best natural control is plentiful bird life. Other chewing pests include grasshoppers and katydids, those bright-green, well-camouflaged, flying bugs. Stick insects are also chewers, but fortunately don't often bother hibiscus. Excelling at camouflage on the bush, the females lay eggs which drop to the ground and emerge as nymphs in spring, climbing up the stem to feed on new leaves. A simple way to prevent them getting there is to band the base of the trunk or stem with a ring of grease – try Vaseline – that stops them in their tracks.

Sucking insects

Sucking insects may be less conspicuous. Although the green vegetable bug and its cousin with the colourful mosaic back are often easy to spot, most sucking insects are a lot smaller. Scale are only 1-2mm long; spider mites, pin-prick-sized pests, can cover the undersides of leaves; powdery or fluffy mealy bugs are a menace, and there's always the common aphid. Be alert for these. Evidence of their presence is white spot or chlorosis of the leaves (spider mites), misshapen buds (aphids), or sooty mould (by-product of mealy bugs and aphids).

To avoid spraying other than a last resort, try first

Pests of the tropics: while gardeners in cooler climates might be alarmed at these creatures, they belong in Queensland, Australia, where this picture was taken. The seedhead chewers do less visible damage than the leaf-eating grasshopper, which in turn is less of a menace than the Queensland hibiscus beetle.

squashing these pests by hand or removing them with a strong-pressured hose. Squashing or hose-blasting mealy bugs will probably not work though, because they coat themselves and the surrounding surface with a waxy substance that resists dislodgement. For mealy bugs, try a paintbrush application of water mixed half-and-half with methylated spirits. Spot applications of this sort (i.e. using a paintbrush) are always environmentally safer than wide-coverage spraying.

Thrips can be difficult to deter as they are fast movers. They're worst in dry conditions, and are not usually a problem in moist areas. Frequent hosing and misting can eliminate them. Whitefly infestation most often occurs in heated glasshouses. It shouldn't be a problem for the home garden where hibiscus have good air circulation.

Other pests

There are other pests of course, most notably, perhaps, the dreaded hibiscus beetle (*Macroura concolor*) – scourge of hibiscus gardens in Queensland, Australia, to which it is fortunately confined.

The beetle is a pollen feeder and chews holes in petals. By boring into the developing bud, both bud and flower are damaged. The safest way of dealing with them is to pick off all flowers and all buds, including those on the ground, and then to burn or boil them to kill both insect and larvae, and to repeat the procedure with successive buds. Flowers are sacrificed but the method can be successful.

For dealing with these destructive creatures there is often no single, safe solution, and unsafe, toxic controls can do more harm than good. But climate can play a role. If you garden in warm or tropical climates, while you may never have to worry about protection from chilling polar storms, insect pests are in your garden year-round. In cool climates insect populations are killed off by frost, which also aids garden hygiene.

Natural controls and non-toxic pesticides

For most of the above mentioned insects, natural controls should be tried. Among natural predators are hoverflies, which can consume huge numbers of aphids and also eat mites and scale insects; praying mantids, which are fully carnivorous and eat aphids, leaf hoppers, caterpillars, beetles and sometimes moths; spiders, which besides consuming flies feed on the larvae of moths and whiteflies; and, best of all, the ever-friendly ladybird. Ladybirds ('ladybug' in North America) should be encouraged everywhere.

Upon hatching, their larvae start immediately to eat aphids, mealy bugs, woolly aphids, thrips and other sap suckers, and they continue into adulthood to clean up these menaces.

Non-toxic sprays should be used with caution, not because of harm to humans but because indiscriminate use can kill both the targeted insect and its predators. The non-toxic preparations described here are short-lived and break down quickly, so they should be used up quickly. Pyrethrum (make your own from *Tanacetum cinerariifolium*, or buy the ready-made produce from garden centres) is a broad-spectrum insecticide. Garlic spray (soak 85 g (3 oz) of whole unpeeled garlic in mineral oil for 24 hours and add 600 ml (1 pint) of water in which a bar of soap has been dissolved; strain, dilute by five times, and spray) will work if it is eaten – it's not a contact spray – and can kill **stink bugs** and **the mosaic-backed vegetable bug**. Chilli spray (1 cup of fresh chilli pulped with ¹/₂ cup of dried chilli and 1 cup of water) controls **caterpillars and loopers**, which are also deterred by sprinkled pepper. Buttermilk spray (¹/₄ cup of buttermilk and 2 cups of flour mixed with 5 litres (1 gallon) of water) will control **mites** when applied to the undersides of leaves every two days. Onion spray can be good for **scale**, **thrips**, **aphids** and **mites** (pour 500 ml (³/₄ pint) boiling water over 1 kg (2¹/₄ lbs) of chopped unpeeled onions and strain; dilute 20 times with water for spraying). Applications of ash from the fireplace, provided it's wood ash and dry, can control leaf-eating **earwigs**, **grasshoppers** and **katydids**. Some gardeners recommend diluted Jeyes fluid (antiseptic) as a good general insecticide.

While chemical insecticides which are systemic (such as Malathion or Maldison) can cause complete defoliation and are to be avoided, milder non-toxic systemics, such as neem oil, can be tried. This oil, widely used in India for a host of purposes, including keeping weevils from food stores, can be made from the leaves of the neem tree. As most hibiscus growers are unlikely to have a neem tree at hand, you can buy the commercial oil and dilute it at a rate of one part oil to 200 parts (warm) water. The diluted solution does not store, so use it at once. It's safe to spray on sucking and plant-feeding pests because it doesn't harm the beneficial predators – or humans – and if applied early enough can suspend the egg-laying-to-larvae cycle of most insect pests.

Be careful with all the above mixtures. Although they use safe substances, they should be made up in special utensils kept separate from those used for food preparation, well labelled, and stored apart from kitchen ingredients.

You can also try traps. Cut a piece of white or yellow cardboard the size of your hand, and coat it with Vaseline. Nail it to a stake or tie to a hibiscus branch. Insect pests, especially night-flying ones, will be lured to it, as they are to all light, and will stick fast. Such a simple device is good for assessing what's in your local insect population. Try it and you'll be surprised at the sticky collection. A 'trap' with a different purpose, that of catching **aphids**, can be made from a liquid containing sugar and yeast.

Another non-toxic suggestion is diluted Marmite laid around the garden in small quantities to attract predators such as hoverflies and lacewings, both of which feed on aphids.

A recipe for 'The Good Oil' from Hibiscus World in Queensland, Australia, uses equal-parts vegetable oil and liquid dishwashing detergent, diluted at the rate of 10 ml (2 tsps) to 1 litre (4 cups) of water. It is reportedly effective both at killing insects and repelling them.

Be aware that none of these sprays or strategies will completely eliminate your plant-damaging insects, but eradication should not be your objective. Aim instead to manage garden problems by keeping a balance of some pests and some predators. (If you did succeed in exterminating all the damaging insects which are the food of their predators, the natural balance would be upset, predators would keep away, and so the next infestation of pests would reach plague proportions.)

On the subject of predators, feline predators add to pest problems by keeping birds out of gardens. Insectivorous birds actually consume huge numbers of pests if given the chance. It may be possible to train cats not to catch birds if disciplined from an early age, but really, all cats catch birds. A cat-free garden is often the most pest-free.

Hibiscus diseases

Phytophthora, the soil-borne wilt disease associated with soils that dry out in summer and are wet in winter, can kill many ornamental shrubs, including hibiscus. It pays to avoid the kind of conditions favourable to *phytophthora* because affected plants usually do not recover. The problem rarely occurs where hibiscus are given good drainage and ample compost.

'Collar rot', visible as a browning discolouration, often slimy, at the root join, is prevalent where winters are cool and damp, and drainage poor. You can try a treatment of removing all affected stems and bark and painting the discoloured area with Bordeaux paste. Clear the surrounding ground completely and spray with a strong garlic solution. **Sclerotina rot** affects stems at ground level. Repeated applications of strong garlic or chamomile tea can help.

If your hibiscus has succumbed to rot and you want to replant in the same spot – which is inadvisable but unfortunately sometimes unavoidable in a small garden – you must first improve the site. Start by removing the affected plant and burning it, then dig up the whole area beyond root depth and spread, distributing the infected soil in an unimportant area, such as plugging a hole in the drive. Some gardeners advocate lighting a fire and spreading it over the entire site before refilling, as fire will kill off any remaining pathogens. The next step is to infill for drainage by lining the cavity with a layer of scoria or pumice over coarser stones, adding sand, and then filling the cavity with new topsoil. You can be extra cautious and use sterilised potting mix.

Besides diseases of the soil, there are fungal diseases that can affect foliage. **Leaf spot** is the main one to watch for. Dark brown or black spots on leaves indicate the presence of pathogens which are most active in wet weather. Heavily affected spotting, where the individual spots have merged to discolour the entire leaf, will result in shedding, and shrubs so depleted can probably only be salvaged by the application of a fungicidal spray, such as Bordeaux. Liquid Bordeaux can be bought from garden retailers or made at home to an exacting recipe of copper sulphate and calcium hydroxide. A recipe for Bordeaux paste uses the same two ingredients less diluted, together with skim-milk powder for viscosity.

In general, compost and mulch are the best root-rot preventatives. Dolomite sprinkled on the area can also help. An all-purpose spray for fungal prevention can be made from a 'weak tea' of nettles, comfrey, yarrow or horseradish. Even better, a weak liquid poured off from seaweed soaked in fresh water has a four-fold benefit: it's a foliar fertiliser, it promotes frost-resistance, it's an insect repellent and a fungicide, and the residual seaweed can be composted or used as a mulch.

Chemical control

Gardeners who choose to use toxins are strongly urged to use them only as a last resort – and sparingly. Never spray in sunlight or in the heat of day. Insecticides containing the chemical Carbaryl (trade names include 'Bugmaster') and Diazinon are considered effective, while fungicides containing Triforine (trade name 'Saprol') and Copper oxychloride are also used in the nursery industry. (Avoid Benomyl – 'Benlate' – as it is one of the most hazardous of all.) Never combine insecticides with fungicides as the combination can be toxic to hibiscus as well as humans. Protective clothing and equipment use are your responsibility. The possibility or risk of any spray drift is likewise your responsibility, and in most Western countries this is now upheld by law.

Eight rules of garden hygiene

· Monitor your plants
· Remove and burn diseased or infected leaves or plants
· Prune out dead wood
· Pick fallen or rotting fruit from fruit trees
· Keep out pollutants from neighbours, roads and traffic
· Do not use toxic pesticides, and use organic ones with discrimination
· Encourage birds
· Compost

Solutions to other problems

Sparse foliage
Apart from insect infestation or the general poor conditions that result from undernourishment, the most frequent cause of sparse foliage is insufficient sun. Site hibiscus in maximum sun.

Crinkled leaves
Check for aphids or leaf hoppers on the undersides of leaves. Large colonies of aphids can suck the life out of leaves before they are noticed. Remove the affected leaves along with the offending insects. If the infestation is severe, try spraying.

Deformed and sickly foliage
If it is possible or likely that your hibiscus has been exposed to systemic insecticides, it will quickly sicken. Hibiscus do not tolerate systemic chemicals,

such as 'Maldison', which are absorbed by the plant. Good feeding and watering may help it recover.

Yellow leaves
Yellowing of leaves at the base of branches and stems occurs naturally as the hibiscus sheds old foliage. If yellowing occurs higher up the stem, especially at the growing tips, it could indicate magnesium or iron deficiency – trace elements easily added to the soil. Alternatively, the discolouration could be a sign of overwatering or overfeeding with chemical fertilisers, in which case suspend both. Yellowing is another sign of exposure to systemic insecticides.

Yellow leaves with green veins
Yellow leaves with green veins are a clear indication of iron deficiency. Apply a good feeding of compost and manure, and scratch in freshly collected seaweed, if you can get it. Iron chelate is often recommended specifically to target the deficiency. Dig this in lightly – never pack down tightly round the stem. Mulch, kept clear from the base of the stem, can also help.

Other leaf discolouration
Brown margins might be windburn, in which case watering and light feeding can restore health. Persistent and prolonged windburn, indicative of a windy climate to which hybrids may not be suited, is a problem that might be best solved by replacing hybrids with wind-tolerant coastal species, such as those described in Chapter 4.

Salt toxicity also shows up as brown leaf margins. Treat with generous watering. Leaves that turn brown at the tips, purple at the edges, and become brittle could indicate a lack of potassium.

Frost damage
Once frost has damaged your hibiscus, there's nothing you can do. If severe, the plant may not recover at all; if light, it will struggle back of its own accord to produce new growth in spring. *Do not*

prune affected parts until the weather has warmed and new growth is established. If you live where frosts can be expected every winter, you may consider pot culture to enable you to move the plant to shelter each year.

Weeds

Any weeds growing around a hibiscus are competing with it for nutrients. Remove them. Remember that the feeding roots of hibiscus are very close to the surface, depending on the same top few centimetres of soil occupied by weeds. The easiest way to keep the area weed-free is to thoroughly clear it before covering with a layer of mulch; the mulch will keep the weeds at bay.

Lawn encroachment

When hibiscus are sited in a lawn, an area of about 1 square metre (3 square feet) should be kept clear of competing growth. Mulching will deter the grass from encroaching again.

Lichen

Grey, green or silver lichens occasionally establish themselves on old mature trees. They won't bother the hibiscus but may bother the gardener. (Some gardeners love and encourage lichens.) It would have to cover and smother the whole shrub before harming it. However, if you want to remove the growth, scrub it off with a scrubbing brush or apply Bordeaux paste in winter.

Ants

Many ant species are valuable predators. They only invade hibiscus if there's something there for them to feed on, and although the odd ant may be attracted to nectar, ants in large numbers are usually after the honeydew excreted by aphids or mealy bugs. Deal with the aphids and the ants will go too. Moisture is an ant deterrent – they increase in dry places – as is pyrethrum spray. Banding the base of the hibiscus stem with Vaseline will prevent ants from travelling further up.

Poor flowering

Few flowers in an otherwise healthy plant may result from too much shade, or, if sun is adequate, from excessive nitrogenous feeding which promotes leaf growth at the expense of flower production. Bud damage may also be a cause.

Buds but no flowers

If your hibiscus are under attack from borers or beetles penetrating the buds before they open, the ground will show the evidence. The treatment of removing all buds, both from the bush (including the tips of branches) and from the ground below, and burning them, followed by insecticidal spraying, should reduce the problem, but may not eliminate it.

Bud drop can also result from lack of water during warm sunny weather; it is important not to let hibiscus dry out.

Flowers appear late in the season

The most likely cause of late blooming is sun and temperature – not enough sun and not warm-enough temperatures.

Blooms change colour

The hybrids sold today are the result of years of intensive hybridising, of crossing one with another, and often backcrossing again to achieve desirable attributes. As with many such 'man-made' garden plants, dominant features of parent plants or earlier ancestry occasionally appear after they have been bred out. Sometimes hibiscus revert to colours of an earlier form; sometimes they'll revert from double to single, from ruffled to unruffled, or from overlapped or windmill single to plain single. This happens most often out of season, while true-to-type blooms can be expected in summer and autumn. If the problem persists, you could consider asking your supplier to replace the plant with another whose performance matches the description.

Above: 'Fire Engine'
Below: 'Lady Cilento'

Above: 'Fiesta'
Below: '442 Battalion'

Above: 'Sylvia Goodman'
Below: 'Ben Lexcen'

Above: 'Thora'
Below: 'June's Joy'

CHAPTER 8

Propagation

Species hibiscus can be grown from seed and the offspring will be true to type. The species described earlier can be reliably propagated in this way (some of them self-seed prolifically), although the time from seed-sewing to a shapely bush in flower is a lot longer than the faster method of taking cuttings.

Hybrids, as distinct from species, cannot be propagated by seed to produce identical offspring. This is because the parental complexity of hybrids results in variable seed, with some seed reverting to one parent and others to an earlier ancestor, and so on, so it's possible for new progeny to resemble only slightly their seed-donor parents.

Propagating species by seed

Germination may be hastened by scoring the surface of the seed with a razor blade, but the procedure is not essential. Use commercial seed-raising mix in a shallow container. Press one or two seeds into the mix to about the same depth of the seed itself, and water well. Keep out of the sun in a place of even temperature. After about three weeks a new seedling, or two, will appear and shoots will develop. Once it has reached about 20 cm (8 in) high you can pot it up, and flowering will be about another five to twelve months away, possibly longer.

Other propagation methods

Other methods of propagation include grafting, air-layering and tissue culture. These techniques call for

some expertise, and most amateur gardeners would prefer to leave these to the specialists. Tissue culture, for example, the modern method of mass production, can only be done in laboratories. For most home gardeners who have admired the hibiscus of another gardener and want to acquire it, or whose own hibiscus is so pleasing they want more, or who are moving homes and can't take a large shrub with them, fortunately there's a simple method of reproduction: taking cuttings.

Taking cuttings

In late winter or early spring, just before the shrub begins its first warmer-weather flush, select a hardwood stem for cutting. Nurserymen intent on production yields may take cuttings of softwood tips, particularly if parental material is in short supply, or medium wood, or hardwood, or all three of these, but the home gardener will find hardwood cuttings easiest to handle. Select the best stem you can find: straight, smooth, as thick as a pencil or thicker, and unblemished.

Now, using scrupulously clean secateurs, cut at the node (i.e. where leaves or buds join the stem) at an angle of about 45°, and then trim to about 12 – 15 cm (5 – 6 in) long; professionals use a sharp knife for this task. Remove any leaves. Place the cuttings in a pot of coarse material, such as sand, perlite, or a mix of these, and include a little peat – about one part sand to one part peat. Don't push the stems deep into the medium – 3 cm (1 in) is plenty. Bottom heat is not necessary, nor is hormone rooting powder. (Experiments have shown that the

'Patricia Noble'

difference between applying hormone preparation and not, is about 20 per cent; commercial growers aim for a 100 per cent strike, but most home gardeners are probably happy with 60 – 70 per cent.)

Place a hoop of wire over the pot by pushing each end of the wire down the side, to create a support for the plastic covering you should then attach. Position the pot in a warm spot away from direct sun. Warmth is important, even if bottom heat isn't (of course, commercial propagators always use bottom heat), so if the climate is cool, wait until the weather heats up. If the medium was well watered at the start, further watering shouldn't be necessary, so you can forget all about the cutting and go forth into the garden. If you forget for long enough, you'll be agreeably surprised at seeing your cuttings when they've struck. This will take about six to eight weeks if an optimum temperature of 20° to 25° C is maintained, and longer in cooler temperatures. Let

Opposite Top: 'Flower Girl'

Bottom: 'Molly Cummings'

them develop more green growth (another month or so), and when they start to really shoot away, put each one into its own individual pot. Should the strike be so successful and vigorous that the cuttings' fine roots are enmeshed, separate them gently by immersing first in clean water. Pot into a mix with some nutrition in it, using such ingredients as compost, peat, commercially bagged potting mix, or quality loam with well-rotted manure – together with a generous scoop or two of coarse sand for drainage – and water well.

It's in these individual pots that your new hibiscus shrubs will start to develop. Keep them out of direct sun and watch over them until you feel they've grown enough to be ready to plant in the garden, or into the larger containers you intend keeping them in. Before their final move outside, expose them gradually to more and stronger sunshine over a period of several weeks. Remember that slugs and snails are never interested in mature leaves – they much prefer the tasty new leaves of an infant plant, so be vigilant and you'll be rewarded.

List of hibiscus hybrids

The following hybrids are a selection from the thousands that are grown and sold around the world. It is impossible to name them all. Simultaneous or parallel hybridising in different countries has sometimes resulted in the same plant having different names. This doubling-up is becoming more apparent today because of the increase in import-export mobility, as plant material from which to propagate, as well as live seedlings, are now transported by the thousand around the globe. While the name confusion is a disadvantage, the big advantage for the home gardener is that this increase in trade is increasing the range of plants available for sale.

The following list is a selection from among the most popular named hybrids. While it makes no claim to be comprehensive (and may also be seen as more representative of Australia and New Zealand than elsewhere), the list is likely to include many that are widely available. As a very general guide, the symbols A, B, C and D have been used to indicate the countries where each hybrid is known to be available for sale, or where it is known or recorded as growing locally – or both. The symbols do not signify country of origin.

A = Australia
B = New Zealand
C = UK/Europe
D = USA

Opposite: 'Annie Wood'

Above right: 'Agnes Galt'

Note also that a hybrid bred and sold in one country is often grafted onto rootstock originating in another, and that there are many hybrids included here that can only be grown successfully as grafted plants.

As for descriptions of these hybrids, the details here are brief. (See also classifying descriptions in Chapter 2, Cultivation.) 'Hardy' means a little more cold tolerant than most. Bush classification refers to height only (low – up to 1m (3 ft); medium – about 1m (3 ft) to over 2 m (6 ft); tall – over 2.2m (7 ft)) and does not take into account shape, spread, or growth habit. Nor do they describe flower details, such as tufted or overlapping windmill petals, or consider which hybrids grow on their own roots and which can only be grown as grafted plants, and so on.

The list should therefore be consulted in conjunction with nursery catalogues which usually provide fuller details, or, better still, with consulting knowledgeable staff. In every country, the best source of reliable plants, and of advice in choosing them, is a local supplier.

ABLAZE	B	single crimson-orange, yellow edges and dark eye, medium bush
AGNES GALT	B	single deep satiny pink, vigorous, medium bush, hardy
ALBO-LASCINATUS	A	single soft pink with darker eye, small flower, very tall bush, hardy
ALL AGLOW	A	single orange-apricot with yellow blotches, low bush, slow growing
ALOHA	AD	single cerise-pink with wide lemon-gold edge, medium bush
AMBER SUZANNE	ACD	double pink and white long-lasting blooms, medium bush
ANDERSONII	A	single scarlet pendulous flower, tall, bronze foliage, good as hedge
ANNA ELIZABETH	AC	double pink and cream, medium bush
ANNIE WOOD	ACD	single cream and yellow with red eye, medium bush
ANVIL SPARKS	B	single vivid orange, medium bush
APACHE	AD	single orange with red eye, medium bush
APRICOT QUEEN	B	single apricot with red eye, medium bush
APRICOT PARADE	B	double apricot with red eye and lavender halo, medium bush
AUTUMN TIME	AC	single orange and red long-lasting blooms, medium bush
BEACH BALL	ACD	double gold long-lasting blooms, medium bush
BEN JAMES	B	single light red, medium bush, prolific
BEN LEXCEN	ACD	single apricot with pink eye, medium bush
BERRIED TREASURE	AD	single lavender long-lasting blooms tall bush
BIG APPLE	ACD	semi-double deep red, medium bush, prolific
BIG TANGO	ACD	single tomato-red with paler eye, low bush
BLACK KNIGHT	ACD	single dark red with cream spots and splashes, medium bush
BLACK MAGIC	ACD	single deep rose with black eye, long-lasting blooms, medium bush
BLAZEAWAY	ACD	single red edged with orange, medium bush
BLUEBERRY TART	AD	single lavender miniature flowers, weak spindly growth
BLUSHING BEAUTY	B	double cream with crimson eye and pink veins, ruffled, medium bush
BONFIRE	AC	single orange-red, low bush
BRIGHT EYES	B	single ochre-yellow with red eye, medium bush
BRIGHT LIGHTS	B	single deep vivid orange-red, medium bush
BRONCO	AD	single yellow with red eye, medium bush
BRUCEII	ACD	single yellow with white eye, tall bush, vigorous and prolific
BUMBLE BEE	ACD	double orange and red, medium bush
BUTTERFULY WINGS	A	double cream or pale apricot, medium bush (syn. PEARL HARBOUR)

Above: 'Berried Treasure'
Below: 'Anvil Sparks'

Above: 'Butterfly Wings'
Below: 'Big Tango'

CALYPSO DANCER	A	semi-double, apricot, ruffled blooms, tall bush
CAMEO QUEEN	AB	single creamy lemon with pale pink eye, low-medium bush, prolific
CANDY	A	single pink edged with white, medium bush, prolific
CANE FIRE	A	single orange with yellow spots, medium bush
CARDINAL	AB	single red, medium bush, hardy (syn. BUCCANEER)
CARMELITA	AC	single ruffled red, medium strongly shaped bush
CARNATION	ACD	double dark red, small picot-edged flower, low bush
CARNIVAL QUEEN	AD	single ruffled, red with cream splashes, medium bush
CASTLE WHITE	A	single white, sometimes tufted, medium bush, minimal pruning
CATAVKI	A	single deep claret-red, medium bush, prolific
CERES	AD	double gold, small flower, medium bush
CHANTAI	A	single rose red, medium bush
CHERIE RITCHIE	A	single red-gold, low bush, good container plant
CHESTER FROWE	AD	single orange-red, medium bush
CHIMARIE	A	double pink-cream-white, medium bush, prolific
CINDERELLA	A	single deep pink and red, low bush (syn. BURGANDY BLUSH)
CINDY	A	single deep pink and cerise, medium bush, prolific
CLAIRE MCLEOD	A	single red splashed with white, medium bush
CLAN MCGILVARY	A	single orange, medium bush, fast growing on own roots
CLOUD DANCER	ACD	semi-double, deep rose-pink, medium bush
COCONUT ICE	ACD	single pink with white spots, medium bush, must be grafted
COLOMBUS	ACD	single apricot, medium bush, shiny foliage, prolific
COMEDY KING	A	single red with yellow veins, medium bush
CONQUEROR	B	single buff small flowers, medium bush, prolific
CONSTANCE	A	double pure white, crepe textured, medium bush
COPENHAGEN	ACD	single orange-red, medium bush
COVAKANIC	A	single multi-coloured, dark eye, tall bush
CREOLE FLAME	ACD	semi-double small brown flower, low bush
CROMWELL	B	single creamy white with pastel pink eye, medium bush (syn. THE BRIDE)
CROWN OF BOHEMIA	ABCD	double gold, medium bush, hardy, good container plant
CUBAN VARIETY	ACD	single apricot with red eye and veining, medium-tall bush, prolific
CULTURED PEARL	ACD	semi-double pink, tall bush
D J O'BRIEN	B	double dark apricot with carmine eye, medium bush, hardy
D J O'BRIEN YELLOW	B	double yellow form of above
DADDY'S GIRL	AD	single lavender, medium bush
DAWN	A	single soft pink with red eye, tall bush, prolific
DAZZLER	B	single bright red and yellow, medium bush
DESERT DAWN	A	single brown with pink eye, medium bush
DEVIL'S GOLD	AD	single very textured gold with brown splashes, red eye, low bush
DOROTHY	ACD	single cream with red eye, low bush (syn. MADONNA, RITA)
DOROTHY BRADY	AC	semi-double rose-pink, medium bush, vigorous and prolific
DOROTHY OLIVE	AC	double orange with red eye, medium bush

'That's Pink'   'Eden Rose'

DOUBLE JOHNSONII	A	double apricot with claret eye, medium bush, prolific
DOUBLE RHONDA D	A	double orange with gold edge, tall bush
DUA	A	semi-double ruffled rose, medium bush
DUSKY BEAU	ACD	single ruffled silvery lavender with plum eye, medium bush
ECLIPSE	A	single deep brown with red eye and pink veins, medium bush
EDEN ROSE	B	single cerise-pink, ruffled, compact medium bush
EDNA MORRIS	A	single red with creamy yellow spots, tall bush
EL CAPITOLIO	AD	single fluted bright scarlet miniature flower, low bush
ENID LEWIS	A	double pale pink, tall bush, prolific
ENTERPRISE	A	single gold with dark red eye, tall bush
EURILLA SUNSHINE	AD	single yellow with red eye, tall bush
EVELYN HOWARD	A	single orange with red eye, large textured flower, medium bush
EXPO	AC	single red splashed and spotted with white, medium bush
FANFARE	AC	single very deep red with almost black eye, medium bush
FANG	A	single red with white eye, small flower medium bush
FANTASIA PINK	ACD	single satiny pink small flower, tall bush, prolific
FANTASIA WHITE	ACD	single white small flower, tall bush, prolific
FARRINGTON	A	single gold with red eye, low bush, good pot plant
FELICE	AC	semi-double rose-pink edged with salmon, medium bush

FIESTA	AC	single ruffled orange with pink eye, medium bush
FIFTH DIMENSION	ACD	multi-coloured: red eye, white veins, beige to orange edge, gun-metal grey, medium bush
FIJI ISLAND	AD	single deep pink with darker eye, small flower, tall bush, vigorous
FIJIAN PINK	AB	single pink with red eye, medium bush
FIJIAN WHITE	AB	single white with red eye, medium bush
FIRE ENGINE	AD	single bright orange-red, medium bush
FIRE DANCE	A	semi-double orange with near-black eye, ruffled, medium bush
FIREFLY	A	single yellow with pale pink eye, low bush, hardy
FIRE TRUCK	A	single orange-red with red eye, medium bush
FIERY LIGHT	A	single orange with pink eye, tall bush
FLAMINGO STAR	ACD	double pink, large flower, medium bush
FLETCHER	A	single chocolate brown with red eye and veins, low bush
FLORIDA SUNSET	ACD	single orange with red eye, small flower, tall bush
FLOWER GIRL	AC	single bright pink, medium bush, prolific, easy
FOR PETE'S SAKE	AD	single lavender with red eye, medium bush
FLYING DUTCHMAN	ACD	single apricot-orange with white eye, medium bush
FOSTER BRADY	AD	semi-double pink and cream with red eye, tall bush
442 BATTALION	ABCD	single khaki with pale mauve eye, upright medium bush, prolific
FRANK ALAND	A	single red with dark red eye, medium bush
FRANK'S LADY	ACD	single pink with gold edge and white veins, medium bush
FRED KERTZE	A	single red with yellow edge, ruffled, medium bush
FREDDIE BRUBAKER	AD	single gold with red eye, large crepe bloom, low bush, prolific
FULL MOON	ACD	double yellow, tall bush, prolific
FUTURAMA	AD	double apricot and lavender, medium bush
GALAXY	A	single gold with carmine eye, medium bush
GANMOR GLORIANA	A	semi-double yellow with pink eye, tall bush
GARDEN GLORY	B	single light pink, large flower, medium bush
GEM	B	single brilliant orange, large flower, medium bush
GENERAL CORTEGES	ABCD	single scarlet red, tall bush, attractive foliage, hardy
GLORIA	AD	single orange, small flower, tall bush
GOLD	ACD	single bright yellow-gold, crepe flower, low bush, glossy foliage
GOLD DUST	AD	single golden apricot with cream splashes, ruffled, medium bush
GOLD COAST CITY	AD	double orange large flower, medium bush
GOLD SPLASH	A	single yellow-gold with red eye, medium bush
GOLDEN BELLE	ABD	single yellow large flower, low bush
GOLDEN HONEY	B	single-to-semi-double bright yellow with white centre, medium bush
GOLDEN PLATTER	ACD	single gold with pale pink eye and white halo, low bush
GOLDEN QUEEN	B	single deep yellow with crimson centre, medium bush (syn. GOLDEN ORIEL)
GOLDEN SUN	B	single golden-orange with red eye, medium bush
GRAHAM MCGILVARY	A	single red, ruffled, medium bush
GRANNY'S BONNET	AD	double lavender multi-coloured large blooms, low bush

'Freddie Brubaker'

GREAT SATAN	AD	single apricot-brown with yellow splashes and red eye, low bush
GREAT WHITE	AD	single large creamy white, tall bush
GWEN MARY	AD	single pink, medium bush
HAPPY NEW YEAR	A	semi-double tan with red eye, medium bush
HAROLD HOLT	A	single purple-red with deep red eye, medium bush
HARRY HARPER	A	single deep crimson very ruffled flower, medium bush
HARVEST GOLD	B	single gold with purplish-red eye, large flower, low bush
HARVEST MOON	AD	semi-double yellow large flower, medium bush
HAWAIIAN RED	AD	single deep red, ruffled, medium bush
HAWAIIAN SKIES	A	single cerise-pink, tall bush
HAWAIIAN SUNSET	A	single pink with gold edge, medium bush
HAZEL CUMMING	A	single pink with pale lemon edge, medium bush
HERM GELLER	ABCD	single honey-brown and gold with red eye, medium bush
HIGH RUFFLES	A	single gold ruffled, low bush
HOLLY'S PRIDE	A	double apricot large flower, very tall bush
HONEY DO	ACD	single gold large flower, medium bush
HONEY GOLD	A	double yellow-gold with white eye, medium bush
HOT MUSTARD	A	single unusual gold-ochre tan, ruffled, medium bush
HOT PEPPER	A	single orange-red, tall bush
HOT SHOT	AD	single red and yellow, ruffled, medium bush
HUMMING BIRD	ACD	single pink with red veins, small flower, low bush
ICELAND GIRL	ACD	semi-double white with red eye, medium bush
INDIAN MAID	ACD	single orange with creamy yellow patches, medium bush

INSIGNUS	*ACD*	single gold with red eye and white halo, medium bush
ISABELLE BARLING	*ACD*	single lavender-pink and deep pink veins, medium bush, prolific
ISLAND EMPRESS	*ACD*	double cerise-pink with burgundy centre, tall bush
ISOBEL BEARD	*ABCD*	semi-double lavender with red eye, medium bush, prolific
JACK CLARK	*B*	double bright orange large flower, medium bush
JALMA	*AD*	single pink with red eye, medium bush
JAYELLA	*AD*	single gold and tan large flower with red eye, tall bush
JAY'S ORANGE	*A*	single orange, low bush
JERRY SMITH	*ABCD*	double rich yellow, low bush
JESSIE LUM	*A*	single pink with red eye, ruffled, medium bush
JIM HOWIE	*A*	single orange and yellow, ruffled, medium bush
JOANNE BOULIN	*A*	semi-double pink cream and lemon, medium bush
JOCKEY'S JACKET	*A*	single red and gold, medium bush
JOL WRIGHT	*ACD*	double orange large flower, low bush
JOLLY MILLER	*A*	single orange with yellow splashes, medium bush, prolific
JOYCE A	*A*	single lavender-pink, medium bush
JULIUS CAMFIELD	*A*	single scarlet, tall bush, shiny foliage
JUNE'S JOY	*A*	single red with gold edge, tall bush
KATY D	*AD*	semi-double rose-pink, medium bush
KENMER RHAPSODY	*AD*	semi-double salmon large flower, tall bush
KIN ELLEN	*A*	single pink, medium bush
KINCHEN'S YELLOW	*A*	single yellow with white eye, crepey textured, medium bush
KITTY BEEBE	*AB*	single lavender with red eye and white veins, low bush
KUIMALE	*A*	single orange with red centre, low bush
KYLIE RITCHIE	*AD*	single orange with yellow edge, medium bush
LADY CILENTO	*ACD*	single bright orange with yellow spots and stripes, tall bush
LADY FLO	*A*	single lavender with red eye, medium bush
LAMBERTII	*ACD*	double red, low bush, good container plant
LAVENDER LADY	*ABCD*	single lavender with red eye, medium bush, prolific
LEMON CHIFFON	*ACD*	single lemon-yellow flushed pink with white eye, medium bush
LIGHT FANTASTIC	*B*	double deep pink, large flowers, medium bush
LITTLE YELLA	*AD*	single yellow, small flower, tall bush
LUTINO	*AD*	single yellow with white veins and red eye, tall bush
MAC DELVEON'S YELLOW	*B*	semi-double clear yellow, vigorous upright bush (syn. NEW ZEALAND YELLOW)
MADANG	*A*	single pink with gold edge, small flower, tall bush
MADONNA	*ABD*	single creamy white with deep red eye, low bush (syn. DOROTHY, RITA)
MADELINE CHAMPION	*A*	single orange, tall bush
MARJ DOLLISSON	*ACD*	single pink and cream with scarlet eye, large flower, low bush
MARJORIE CORAL	*AD*	single rose-pink with red eye large flower, medium bush
MAROON STARS	*AD*	single red with yellow-cream blotches and purple eye, tall bush
MARTHA FLEMING	*AB*	double rich blush-red with wide white edge, ruffled, medium bush

'Light Fantastic'   'Madonna'

MARTHA IRENE	A	single yellow with pink eye, tall bush
MARY WALLACE	ABCD	single deep orange with yellow edges, medium bush
MASON RED	A	single orange-red with white splashes, ruffled, medium bush
MASQUERADE	B	double flamboyant brick-red petals edged in orange, medium bush
MEEKE MEEKE	A	double orange, medium bush
MELLIE MAY	AD	single apricot and gold, tall bush
METEOR	AD	single bright yellow with red eye and white veins, low bush
MINI SKIRT	ACD	single cerise-red with white spots and streaks, ruffled, low bush
MISS HAWAII	ACD	semi-double orange, medium bush, slow growing
MISS LIBERTY	ABCD	single cream with pink veins, medium bush
MISS NORTH MIAMI	AD	single red, medium bush
MOLLY CUMMINGS	ABCD	single rich velvety red, medium bush, hardy, reliable and easy
MONET	A	single buff with red eye, large flower, tall bush
MONICA	A	single white, medium bush
MONIQUE MARIA	ACD	double pink with apricot edge, ruffled, medium bush
MOONSHOT	A	single yellow with red eye and white halo, tall bush, prolific
MORNING STAR	ACD	single lavender-pink, medium bush
MOUNT SHASTA	ACD	double creamy white, ruffled, medium bush, glossy foliage
MRS ANDREASSON	A	double lemony cream and pink, tall bush, fast grower
MRS GEORGE DAVIS	A	double rose-pink, ruffled, tall bush, hardy, prolific
MRS HAYWOOD	AD	single rich red, ruffled, medium bush
MYSTIC CHARM	ABCD	single cerise-red with white border, low bush
MYSTIC HUE	ACD	single lavender-mauve, small flower, tall bush
NATHAN CHARLES	B	single purplish-red with red eye crepe flower, medium bush, prolific
NEW IDEA	A	double red large flower, tall bush
NORMA	A	single gold with pinkish-red eye, low bush

'Pro Legato'

NORMAN LEE	ACD	single deep cerise with broad yellow edge, low bush, prolific
NORMAN STEVENS	A	single apricot-orange with white eye, medium bush
OLD FRANKIE	B	single pink crepe flower, low-to-medium bush, good pot plant
OPEN ARMS	A	single pink and apricot, medium bush, shiny foliage
ORANGE GLOW	A	semi-double bright orange, medium bush
ORANGE GROVE	AD	single orange with red eye and pink halo, ruffled, tall bush
ORANGE MAGIC	ACD	single dark orange with apricot edges, medium bush
ORANGE PARADE	AB	double light orange, medium bush
ORANGE PRIDE	A	single orange with deep pink eye, medium bush, prolific
ORANGE TRIUMPH	A	double orange-red, medium bush
PA WALLACE	AD	double red small flower, medium bush
PACIFIC PEARL	ACD	single lilac with red eye and silvery pink sheen, medium bush
PANORAMA	A	double scarlet, medium bush
PATRICIA NOBLE	A	single salmon-pink with cream streaks and spots medium bush
PEACH BLOW	A	semi-double soft pink with red eye, tall bush
PEEPING TOM	AD	single red with orange border, medium bush
PEGGY WALTON	A	single pale lavender with red eye, low bush, prolific
PERFECTION	A	single bright orange, medium bush
PERSEPHONE	A	single lavender-purple with large red eye, medium bush
PINDARI	A	single apricot with pale pink eye, medium bush
PINK CAMEO	ACD	single pink with purple eye, tall bush, vigorous
PINK DAWN	D	semi-double lavender-pink with paler edges, medium bush
PINK GEM	ACD	single pink with red eye, small flower, tall bush
PINK PSYCHE	ACD	single satin pink, small fringed flowers, tall bush
PINK RADIANCE	ACD	single deep pink with paler edges, medium bush, prolific
PINK RAYS	ACD	single pink and lemon with pink veins, medium bush

PINK SATIN	ACD	single pink, medium bush
PRECIOUS NOVA	ACD	single orange with gold edge, medium bush
PRELUDE	A	single deep orange with pink eye, medium bush
PRETTY BABY	A	single lavender-pink, medium bush
PRIDE OF LOCHLEYS	B	double light red, medium bush, hardy and vigorous
PRIMROSE	ACD	single lemon with white eye, medium bush
PRINCE OF ORANGE NZ	B	single bright orange with lemon edge, medium bush, prolific
PRO LEGATO	ABCD	double deep purplish-red, low bush
PSYCHE	ACD	single shiny red small flower, tall bush, hardy and prolific
PURPLE PASSION	ACD	single purple-red with burgundy eye and cream border, low bush
RASPBERRY CREPE	ACD	single pink crepe flower, medium bush
RED BOMB	AD	single red with yellow spots, ruffled, medium bush
RED PARASOL	A	single deep pink-red with cream edge, medium bush, prolific
RED ROBIN	A	single red ruffled, medium bush, prolific
RENA GEORGE	A	single yellow with red eye, medium bush
RHONDA D	A	single hot orange with yellow edge, ruffled, tall bush, fast growing
ROSALIND	A	semi-double orange and red with lemon blotches, tall bush
ROSE SCOTT	ACD	single rose-pink with red eye, tall bush, hardy, good as hedge
ROSS ESTEY	AD	single pink and peach, ruffled, medium bush, prolific
RUBY ROSE	AD	single pink with maroon eye, small flower, low bush
RUTH WILCOX	A	single pink with dark pink eye, small flower, tall bush (syn. ALBO-LASCINATUS)
SABRINA	ACD	double cerise-red, ruffled, tall bush
SAKETANI BLUE	AB	single lavender with pick eye, low bush, will only grow on grafted rootstock
SAMANTHA CHAPPELL	A	single soft rose-pink, low bush

'Ross Estey'

SAN DIEGO BEAUTY	A	single cerise, medium bush, good as hedge
SASSY	AC	double red and yellow, medium bush
SATU	AC	double pinkish-red, tall bush
SCARLET GIANT	AC	single bright scarlet large flower, medium bush
SEMINOLE PINK	AC	single pink with red eye and veins, tall bush
SHARON DEE	AD	single mauve with scarlet eye, tall bush
SHIRLEY HOWIE	ACD	single pink with cream borders, medium bush
SHOCKING PINK	B	single bright rich pink, medium bush
SHOW BUSINESS	AD	single pinkish-red with yellow, medium bush
SHOW GIRL	AD	double red, tall bush
SIBLING X	AD	single red with white splashes, medium bush
SILVER ROSE	ACD	single lavender-pink with silvery eye, crepe flower, medium bush
SIMMONDS RED	B	single tomato red, medium bush
SINENSIS	A	single scarlet red, medium bush, hardy
SLEEPING SINGLE	AD	single red with gold border, ruffled, medium bush
SMOKEY BLUE	AB	single pale lilac with grey sheen, medium bush, not vigorous
SNOW QUEEN	AD	single red small flower, tall bush, striking white variegated leaves
SONATA	A	single salmon-pink with red eye, medium bush, prolific
SPRING SONG	ACD	single orange with pink eye, medium bush, fast growing
SPRINKLE RAIN	ACD	single salmon-pink with cerise eye, small flower, tall bush, prolific
STACEY WINTERS	B	single white with red centre, medium bush
STELLA MCLAIN	A	semi-double red with white edges, medium bush
STORMY DAYS	ACD	double deep lavender fading to silver, medium bush
SUMMER FLASH	AD	single gold-orange, medium bush
SUNDOWN	AD	single deep orange with yellow edges, low bush, prolific
SURFRIDER	ABCD	single orange with red eye, large flower, low bush, prolific
SUSANNAH PHILLIPS	A	single strawberry-pink with lemon veins and edges, medium bush
SUVA BELLE	A	single orange with strong red eye, medium bush
SUVA QUEEN	B	double rose-pink, medium bush, vigorous and hardy
SWAN LAKE	AC	single white, small flower, tall bush (syn. WHITE FANTASIA)
SWEET AMBER	ACD	semi-double cream with red eye, low bush
SYLVIA GOODMAN	AC	single creamy yellow with red eye, medium bush
TAMMY FAYE	ACD	single lavender-pink with red eye and apricot edges, medium bush
TANGO	ACD	single orange, medium bush
TEN THIRTY SEVEN	AD	single orange and red, large flower, tall bush
THAT'S PINK	B	double deep pink, long-lasting flower, medium bush
THE BRIDE	AB	single creamy white with pastel pink eye, medium bush (syn. CROMWELL)
THE PATH	AD	single pink-yellow multi-coloured, low bush, good container plant
THE PEARL	ACD	single white-tinged pale pink, ruffled, medium bush
THELMA BENNELL	A	single cerise-pink, tall bush
THUMBELINA	AC	single cream with red eye and pink edges, low bush, good in pots

'Simmonds Red'

THORA	ACD	single cream and cerise-pink, small flower, tall bush
TOM	A	single orange with yellow edges, medium bush
TOMATO LANI	A	single orange with yellow edges, low bush, prolific
TOPSY	AD	single gold and red, low bush, prolific, good beginners' hybrid
VANILLA SUNDAE	A	single white with red eye, medium bush
VASCO	ACD	single bright lemon with white eye, medium bush
VITAMIN C	ACD	semi-double pale yellow-gold, medium bush
VIVID	B	single orange, medium bush
VIVIENNE	ACD	single deep red, tall bush
WARRIEWOOD GEM	ABC	single brilliant red, medium height, prolific
WHIRLS-N-TWIRLS	AD	single yellow, medium bush
WHITE KALAKAUA	ABCD	double white suffused with pale pink, tall bush (syn. ELEPHANT EAR, PEARL HARBOUR, BUTTERFLY WINGS)
WHITE PICARDY	AB	semi-double pure white, low bush
WILMAE	AD	single orange brown with mauve eye, medium bush
WRIGHTII	A	single white with red eye, tall bush, fast grower, good espaliered
YELLOW FIRE	ACD	single yellow-gold, low bush, good in containers
YVONNE	ACD	single pale apricot with pink eye, tall bush, can be grown from seed
YO-YO	AD	single orange with yellow eye, medium bush
ZELLIE WAEGNER	AD	semi-double lemon and orange with deep red eye, medium bush

Lists of selected hybrids for particular purposes

Reliable beginners' plants

Agnes Galt
Apricot Parade
Annie Wood
Molly Cummings
Old Frankie
Ross Estey
Suva Queen
Topsy

Suitable for containers

Crown of Bohemia
Golden Belle
Madonna (syn. Dorothy, Rita)
Marjorie Coral
Mary Wallace
Nathan Charles
Old Frankie
Surfrider
The Path

Tolerant of cooler conditions

Some of these are used as rootstock for less hardy hybrids

Agnes Galt
Cardinal (syn. Buccaneer)
D J O'Brien
Firefly
General Corteges
Rose Scott
Sinensis
Suva Queen

'Old Frankie'

'Suva Queen'

Dwarf form
Suited to small gardens

Cameo Queen
Granny's Bonnet
Harvest Gold
Meteor
Vitamin C

Long-lasting blooms
These last longer than the 1 – 2 day average bloom

Beach Ball
Berried Treasure
Comedy King
Devil's Gold
Mini Skirt
Miss Liberty
New Idea

Standards

Albo-Lascinatus
Fantasia Pink
Fantasia White
Pink Psyche
Sprinkle Rain
Sylvia Goodman

Hedging

Albo-Lascinatus
Andersonii
Fijian White
General Corteges
Rose Scott

Espalier

Agnes Galt
D J O'Brien
General Corteges
Rose Scott
Ross Estey
White Kalakaua (syn. Pearl Harbour)
Wrightii

Foliage

Andersonii (bronze)
Columbus (glossy)
General Corteges (dark, sometimes variegated)
Kinchen's Yellow (dark, shiny)
Mount Shasta (very glossy)
Snow Queen (striking white variegated)
Warriewood Gem (dark)

Unusual colours
Flower colours are most distinctive when first open, sometimes later changing

Desert Dawn
Devil's Gold
Fifth Dimension
Granny's Bonnet
Hot Mustard
Purple Passion
Wilmae

'Crown of Bohemia'

Fifteen facts

1. The largest hibiscus flower measured was a 30 cm (12 in) diametre *H. moscheutos*, and the longest staminal tube measured was 145 mm (6 in) on the hybrid flower 'Ross Estey'.

2. The oldest hibiscus growing today in Australia is known to have been planted in Camden Garden in 1880.

3. Hibiscus is the fifth most popular plant in cultivation in the world – after roses, azaleas, carnations and orchids, in that order. In Germany the three most popular flowering pot plants are poinsettia, cyclamen and hibiscus.

4. In Asia the hibiscus is called the 'shoeshine' or the 'shoeblack' plant because the flower (*H. rosa-sinensis*) can be rubbed over leather to blacken and shine it. The method is astonishingly effective.

5. The purplish-black juice from crushed red flowers has been used as a dye, as an ink, as colouring for liquors, and to stain paper to make it react like litmus. It was also popularly used by women for tinting their hair and eyebrows.

6. Today the flower is more important to the Asian economy as a dried ingredient in herbal medicine. A recent (1998) order from China, received by an Australian grower, was for dried hibiscus flowers to be exported to China in batches of 25 tonnes!

7. The Chinese also pickle flowers for eating, and use them to make a tea considered to be mildly sedative. Hawaiians eat them raw as an aid to digestion.

8. In Pacific cultures the wearing of a hibiscus bloom over the right ear indicates that the wearer is single and seeks a lover; when worn over the left ear it means the wearer is mated.

9. For floral arrangements, flowers picked from the bush last exactly the same length of time whether or not they are put in water.

10. Hibiscus bark has been used for centuries as a strong fibre for weaving and making cords and rope. Try taking thin strips off a long hibiscus stem to plait into a rope – it can prove to be extremely strong.

11. While the outside fibre was used to make fishing nets, the inside – one of the lightest of all woods – was used as fishing floats. Although cork is used for floats today, many cultures still grow hibiscus for fibre.

12. The species *H. sabdariffa* was once grown as a commercial crop in Australia for producing jam. The fleshy calyces exude an acidic juice, somewhat like redcurrant or cranberry in flavour, which Jamaicans ferment into a drink. The jam recipe: harvest the calyces (i.e. what's left after the flower has wilted, containing the seeds) along with short stalks attached, and cover them with water. Boil for about 45 or 50 minutes. Strain and cool, then measure the liquid and add cup for cup of sugar. Boil until it jells. Cool and bottle.

13. Hibiscus are immune to the herbicide *glyphosate*.

14. One of the most important pollinators of hibiscus flowers is the hummingbird.

15. The number of hibiscus species is still being added to. In 1998 a new species was discovered growing in a confined location in Queensland, Australia.

Index